ON THE GENERATION
OF ANIMALS

MERCER UNIVERSITY PRESS

Endowed by

TOM WATSON BROWN
and
THE WATSON-BROWN FOUNDATION, INC.

A R I S T O T L E

ON THE GENERATION OF ANIMALS

Translated by David Bolotin

MERCER UNIVERSITY PRESS
Macon, Georgia

Mercer University Press, 1501 Mercer University Drive, Macon, Georgia 31207
Copyright © 2025 by David Bolotin

Published in 2026
Manufactured in the United States of America

1 2 3 4 5   30 29 28 27 26

Library of Congress Cataloging-in-Publication Data
LCCN 2025034485

Aristotle.
[De Generatione Animalium. English]
On the Generation of Animals / Aristotle; translated by David Bolotin. Macon,
Georgia: Mercer University Press, 2025.
Includes bibliographical references.
ISBN 978-088146-989-9 (hardback: alk. paper) |
ISBN 978-088146-990-5 (paperback: alk. paper)

Additional Cataloging-in-Publication Data is available from the
Library of Congress.

∞ The paper used in this publication meets the minimum requirements of the
American National Standard for Information Sciences Permanence of Paper for
Printed Materials, ANSI Z39.48–1984.

The typefaces used in this book are Monotype Ehrhardt and Palatino.
Cover image: a medieval folio that includes the beginning of *On the Generation
of Animals*, in the Vatican Library, Vat.gr.1339_0080r_m, by permission of
Biblioteca Apostolica Vaticana, with all rights reserved.
https://digi.vatlib.it/view/MSS_Vat.gr.1339/0080
Cover design by Burt&Burt Studio
Interior design by Susan P. Johnson

Published by
Mercer University Press
www.mupress.org

Contents

Introduction

Aristotle's *On the Generation of Animals* presents itself as an immediate sequel to his treatise *On the Parts of Animals.* For it begins with the assertion that the other parts in the animals have been spoken about, but that there remain the parts contributing to coming into being or generation, about which nothing has been determined. And indeed in *On the Parts of Animals* Aristotle explicitly postpones the treatment of the reproductive organs, along with such substances as semen and milk, promising to discuss them in his speeches about generation (*On the Parts of Animals* 678a22–28; cf. 655b23–27, 689a5–20, et al.). Moreover, the final sentence of that treatise, according to most of the manuscripts that have come down to us, points directly to *On the Generation of Animals* by saying that it remains "next to go through what has to do with the coming into being" of the various animals (*On the Parts of Animals* 697b29–30). And yet despite all this, there is also evidence that *On the Generation of Animals* may belong in a different setting. For that last sentence of *On the Parts of Animals* is absent from some of the manuscripts, in which the text proceeds instead directly to the first sentence of *On the Progression of Animals*, a treatise that examines the parts of animals that are useful for the various modes of locomotion. Admittedly, this seems not to have been Aristotle's own original ordering of these two treatises, since three passages in *On the Parts of Animals* refer to *On the Progression of Animals* as an earlier work (*On the Parts of Animals* 690b15–16, 692a17–18, 696a9–12). But it might indicate a subsequent reordering on his part. And if we allow this as a hypothesis, we can see a different, indirect, path from *On the Parts of Animals* to *On the Generation of Animals*. For the last sentence of *On the Progression of Animals* speaks of it as having followed *On the Parts of Animals*, and it then directs us explicitly to *On Soul*, whose sequel is clearly *On Sense Perception and Sense-Perceptibles*, the first of the treatises that are now known collectively as the *Parva Naturalia* (cf. *On Sense Perception and Sense-Perceptibles* 436a1–4); moreover, the fifth of these treatises,

On Prophecy from Sleep, concludes, according to most of the manuscripts, with an explicit pointer to *On the Motion of Animals*, whose last sentence directs us, again explicitly, to *On the Generation of Animals*. So it appears possible, at any rate, that Aristotle may have intended this more circuitous route from *On the Parts of Animals* to *On the Generation of Animals*, either as a complement to or as a replacement for the direct one.[1] And the view that *On the Generation of Animals* is meant to follow *On the Motion of Animals*, which follows in turn the initial treatises of the *Parva Naturalia*, finds some support in that it contains explicit back references to *On Sense Perception and Sense-Perceptibles* (771b21–22, 781a20–22, 786b23–25, 788a34– b2) and probably also one to *On Sleep and Awakeness*, the third treatise of the *Parva Naturalia* (779a6–7).

There is, however, an immediate difficulty with this view that Aristotle intended *On the Generation of Animals* to follow *On the Motion of Animals*. For in its opening statement, in addition to claiming that the parts contributing to coming into being remain to be treated, he says that while the other causes have been spoken about, the cause that sets in motion— which includes, at any rate, the cause that sets animals in motion—remains, adding that its treatment is in a certain manner the same as that of the coming into being of each (kind of animal). In other words, he explains his project of writing about the generation, or coming into being, of animals as a way of treating the cause that initiates (animal) motion in general, a cause that still remains to be treated. But though the claim that this cause remains to be treated is plausible in a sequel to *On the Parts of Animals*, it seems out of place in a sequel to *On the Motion of Animals*, which is explicitly devoted to the cause, or ruling beginning, of animal motion of whatever sort (*On the Motion of Animals* 698a4–7; cf. 700b9–11). Indeed, in that treatise Aristotle not only investigates this cause at length, but he seems to answer the question of what it is, namely, the good attainable by action, which as an object of thought or imagination awakens the motion of appetite, which in turn sets the animal in motion (*On the Motion of Animals* 700b24–28, 700b35–701a2, 701b33–34). This answer, however, is

---

1 Cf. *Aristote: Les Parties des animaux*, trans. and ed. Pierre Louis, 2nd ed. (Paris: Les Belles Lettres, 1957/2002), x–xi. Also Marwan Rashed, "Agrégat de parties ou *vinculum substantiale?* Sur une hésitation conceptuelle et textuelle du *corpus* aristotélicien," in *Aristote et le mouvement des animaux: Dix études sur le De Motu Animalium*, ed. André Laks and Marwan Rashed (Villeneuve d'Ascq: Presses universitaires du Septentrion, 2004), 185–202.

not in fact Aristotle's last word, for he immediately goes on to raise difficulties with it, as he signals by introducing his extended discussion of how
appetite moves the body with the question of why one only sometimes is
set in motion in accord with one's thought, but sometimes isn't. In the
course of his discussion of this question, which occupies the entire second
half of the treatise, Aristotle makes it clear that one's thoughts are followed
by the intended motions only if the bodily matter that must be acted upon
is of the appropriate amount and character (*On the Motion of Animals*
703b36–704a2; cf. 702a10–21). And this in turn depends on all sorts of
motions and changes coming both from outside and from within the body.
Such causes are all the more obviously decisive, moreover, for the many
involuntary motions, such as sexual arousal, and non-voluntary motions,
such as breathing, that characterize animal life, the discussion of which
Aristotle introduces in the final chapter of the treatise (*On the Motion of
Animals* 703b3–20; cf. *Physics* 253a7–21, 259b6–16).

Thus, *On the Motion of Animals* doesn't give an adequate answer to the
question of what the cause is from which animal motion proceeds, but
leaves us even with the question of what could count as such a cause. And
so it seems possible that Aristotle might have intended *On the Generation
of Animals*, with its promise of a treatment of this cause, as a sequel to that
treatise. Moreover, it is at least consistent with this hypothesis that in his
initial statement that the cause that sets (animals) in motion remains to be
treated, he doesn't say, as he does say with regard to the parts contributing
to coming into being, that nothing has been determined about it earlier.
And as to why the continuation of the inquiry into this cause should take
the form of a study of coming into being, that is a question for the treatise
itself, but I might hazard as an initial suggestion that it makes some sense
in seeking to find a cause or beginning of motion to go back to the earliest
stages of life (cf. *On the Motion of Animals* 700a16–21, b11–14).

Accordingly, I offer this translation of *On the Generation of Animals* as
a supplement to my earlier translation of the *Parva Naturalia* (with *On
the Motion of Animals*),[2] where it would be situated immediately after *On
the Motion of Animals*.[3] I have based my translation on the Greek text of

---

2 *Aristotle: Parva Naturalia with On the Motion of Animals*, trans. David Bolotin (Macon,
   Ga.: Mercer University Press, 2021). See in particular viii–ix.

3 This would place *On the Generation of Animals* before *On Long-Livedness and Short-
   Livedness* and *On Youth and Old Age and Life and Death and Breathing*, the final two
   treatises of the *Parva Naturalia*. And the fact that it contains a reference to a future

On the Generation of Animals edited by H. J. Drossaart Lulofs in the Oxford Classical Texts series.[4] I have not, however, always agreed with his readings. As is standard among editors of ancient Greek works, Drossaart Lulofs includes at the bottom of each page a critical apparatus that reports what he regards as the most significant alternatives to the version that he accepts in his printed text, both readings from different manuscripts and, less commonly, emendations proposed by modern scholars. And I have sometimes based my translation on one or another of these alternate readings. Whenever I have done so, however, I have added a footnote that includes, along with a transliteration of the variant I have chosen, and sometimes my reason or reasons for choosing it, a transliteration and also a translation of the reading in Drossaart Lulofs' text. In making my own editorial decisions, I have been helped by consulting two other editions of the Greek, one with an English translation by A. L. Peck[5] and one with a French translation by Pierre Louis.[6]

In translating I have tried to convey Aristotle's meaning as accurately as I could, while also providing as literal a translation as is compatible with good English. And especially because Aristotle is such a terse and even elliptical writer, the only way I could combine these two aims was to supplement the literal translation with many additions of my own, which I have included within square brackets. These bracketed suggestions, though they are often extremely minor and necessary only to give the translation a normal English style, do however sometimes involve interpretative judgments regarding the passage in question. And though there are so many of these additions that no one could possibly ignore them, those readers who question my judgment regarding this or that passage can at least measure my full translation against the literal version that excludes the bracketed words.

I have accepted the traditional division of this treatise into chapters, even though it is not present in the oldest manuscripts, since it seems rea-

---

discussion in *On Long-Livedness and Short-Livedness* (777b6–8) suggests that it is meant to be followed by these two treatises.

4 *Aristotelis De Generatione Animalium*, ed. H. J. Drossaart Lulofs (Oxford: Clarendon Press, 1965).

5 *Aristotle: Generation of Animals*, trans. and ed. A. L. Peck (Cambridge: Harvard University Press, 1942/1963).

6 *Aristote: De la Génération des animaux*, trans. and ed. Pierre Louis (Paris: Les Belles Lettres, 1961).

sonably intelligent for the most part, and since it would be too jarring for the contemporary reader not to have chapters. But though it is now customary for editors and translators to divide each chapter into paragraphs, I have not done so, since the ease of reading that they provide does not seem to me worth the risk of making inappropriate divisions within Aristotle's presentation of his thought.

David Bolotin
April 2024
Santa Fe, New Mexico

Book One

Chapter 1

Since the other parts in the animals have been spoken about, both in common and separately concerning the ones peculiar to each kind, in what manner each [part] is on account of such a cause, I mean this [cause], the for the sake of something; for four causes have been assumed, the for the sake of which as an end and the articulation¹ of the being² (these ought, really, to be taken pretty much as one) and, third and fourth, the material and [that] from where the ruling beginning³ of the motion [is]—so now [our account]

---

1   The Greek word translated here as "articulation" is *logos*, which can mean a speech or spoken account, as it does below at 715a15–16, but also, as it does here, the rational or intelligible structure brought to light in such an account. Analogously, it can mean both a reckoning, or computation, and also the reckoned amount, or ratio, brought to light in this way. I will generally translate it as "speech," "account," "argument" or "reason," or as "articulation" or "ratio," but where none of these terms seems adequate, I will translate it as best I can and add a transliteration of the Greek in parentheses. Also, since Aristotle later uses a different Greek word that I also translate as "articulation," in the sense of differentiation of the parts of the body, in those few subsequent cases where I use "articulation" as a translation of *logos*, I will add a transliteration of the Greek.

2   The Greek word *ousias*, translated here as "being," is an inflected form of *ousia*, which means primarily an independent being (such as an animal) or one of its parts, as distinct from its qualities or attributes, but which can also mean, as it does here, the "beingness" of such a being, i.e., the answer to the question "What is it?".

3   The Greek word translated as "ruling beginning" is *arkhē*. It can mean both "beginning" and "rule," and Aristotle uses it often to mean a beginning or source, e.g., of a motion, that exercises an ongoing influence or control over it. I will continue to translate it as "ruling beginning," or, in most contexts, simply as "beginning" (a word that I will use only to translate *arkhē* or one of its forms).

concerning the others has been stated, for the articulation (*logos*) and the for the sake of which as an end [are] the same, and the material, for the animals, [is] their parts, for the whole [animal] in its entirety the non-uniform ones, for the non-uniform the uniform, and for these the so-called elements of bodies; but [there is] left, of the parts, those that contribute to coming into being[4] for the animals, about which nothing has been determined earlier, and concerning cause, the [one] that sets in motion, what ruling beginning [it is]. And to examine concerning this [latter] and to [examine] concerning the coming into being of each [animal] are in a certain manner the same; wherefore [our] account has brought [them] together into one, putting these [parts] last in order of the [things said] concerning the parts, and of the [things said or spoken about] concerning coming into being [putting] the beginning next after these. Now among the animals, some come into being from a coupling of female and male, in as many kinds of animals as have the female and the male. For not all have them, but while in the blooded [kinds], in all but a few, this one is male and that one female, [each sex] having been brought to perfection, among the bloodless [kinds] some have the female and the male, so as to generate [offspring] of the same kind, but others generate, yet not [offspring] of the same kind.

---

4   The Greek word *genesin*, translated here as "coming into being," is a form of the word *genesis*, another form of which is translated in the title of this treatise as "generation." Indeed, the word can mean both "generation," in the sense of the engendering or bringing forth of offspring, and also "coming into being," and so in particular that of the offspring themselves. And though Aristotle is clearly referring here to the reproductive organs, the fact that he speaks of their contributing to *genesin* "*for* the animals," which most plausibly means for the *offspring* who receive the benefit from reproductive activity, suggests that the translation "coming into being" is the more appropriate one here. In my view, moreover, the translation "coming into being" is generally the more appropriate one in this treatise, especially since there is also a different word, *gennēsis*, which Aristotle uses to mean "generation" or "generating" exclusively. Accordingly, and to provide consistency in the translation of this important word, I will continue to translate *genesis* as "coming into being" except in those relatively few cases where the context seems to require the translation "generation." And in those cases, I will include after the word "generation" a transliteration of the Greek in parentheses. As for the translation of the title, I have used *On the Generation of Animals*, rather than *On the Coming into Being of Animals*, simply because it is the traditional title.

Of such a sort are as many as come into being, not from animals coupling, but out of rotting earth and residues. But to speak generally, as many of the animals as [are] capable of changing their place, some being able to swim, some to fly, and some to go on foot, by means of their [differing] bodies, all these have the female and the male, not only the blooded ones but some also [that are] bloodless. And of these [latter], some [have them] throughout the whole kind, for instance, the cephalopods and the soft-shelled [animals]; but in the kind consisting of the insects, [only] the majority [have them]. And of these themselves, as many as come into being from coupling of animals of their [own] kind also themselves generate according to their kind; but as many as do not [come into being] from animals, but out of material that is rotting, these do generate, but a different kind, and that which comes into being is neither female nor male. Of such a sort are some of the insects. And this happens reasonably; for with as many as don't come into being from animals, if animals came into being from these coupling, if [they were] like in kind, the coming into being from the beginning of the ones who begat them ought also to have been of such a sort. (And this we claim reasonably; for it manifestly happens in this way in the case of the other animals.) And if [the offspring were] dissimilar, but able to couple, some different nature would in turn come into being from these, and in turn some other from these, and this would proceed to infinity. But nature flees from the infinite; for the infinite is without end, but nature always seeks an end. But as many of the animals as are not capable of locomotion, like the shell-skinned[5] [animals] and those that live by growing on [something], because their being is nearly the same as the plants, just as those don't have the female and the male, neither do these, but they have come to be called [female and male] owing to likeness and analogy; for they have a small differentiation of this sort. For even among plants there are, within the same kind, some trees that are fruit-bearing

25

30

715b

5

10

15

20

---

5   The Greek word translated as "shell-skinned," is *ostrakoderma*, which means literally "with a shell like a potsherd, hard-shelled." Liddell, Scott, and Jones, *Greek-English Lexicon*, 9th ed. (1996). This class includes oysters, clams, mussels, and snails. The other main classes of bloodless animals that Aristotle recognizes are the soft-shelled animals, such as crabs and crayfish, the cephalopods, such as octopuses and squid, and the insects.

and others that do not bear fruit themselves, but that contribute, for those that bear [it], to concocting[6] [it], as happens with the
25   fig and the caprifig. It is the same way also in the case of plants. For some come into being out of a seed, but others as it were from nature acting spontaneously; for they come into being either from the earth rotting or from some parts in plants. For some [kinds] are not themselves constituted separately by themselves, but they come into being on trees [that are] different, for example, the
30   mistletoe. Now concerning plants, they must be examined sepa-
716a rately, themselves by themselves.

Chapter 2

But concerning the coming into being of the others, [the] ani-
mals, [our account] must be stated in accord with the speech that belongs to each of them, connecting [it] with what (pl.) has been
5   said. For, as we have said, someone might posit [as] ruling begin-
nings of coming into being the female and the male most of all, the male as containing the ruling beginning of motion and com-
ing into being, the female as [containing that] of material. And someone might trust this especially from considering how the seed comes into being and from where; for while the [beings] that come into being by nature are constituted out of this, it must not
10   escape notice how it happens that this comes into being from the female and the male. For by this sort of part being secreted from the female and the male, and the secretion being in these and out of these, on account of this the female and the male are ruling beginnings of coming into being. For we call the animal that generates into another, male, and the one [that generates] into itself,
15   female; wherefore also, within the [world as a] whole, people recog-
nize the nature of the earth as female and a mother, and they ad-
dress the heaven and the sun or any of the other such [beings] as

---

6   The Greek word translated as "concocting" is *pettein*, which refers to various transformations, primarily toward maturity or some other healthy or desired condition, that result from heat. These transformations include ripening, cooking (in particular, baking), and digestion. For the sake of consistency, I will continue to translate it as some form of "to concoct," even in contexts where one of these other terms would be more natural.

generators and fathers. The male and the female differ in the account [we give of them] by each being capable of [something] different, and [they differ] to sense perception by certain parts, in
the account [we give], by the male being that which is capable of   20
generating into another, as was said earlier, and the female that
[which is capable of generating] into itself, and out of which that
which is being generated, [while] being present in that which is
generating, comes into being. And since they are distinguished
by [each having] a capacity and by a certain work, and there is a
need of tools for every productive action, and tools for the capacities [are] the parts of the body, [it is] necessary that there be parts   25
both for begetting and for coupling, these [parts] differing from
each other in accord with the way in which the male will differ
from the female. For although it is the whole animal that is spoken
of, the one [as] female and the other [as] male, still it is not female
or male with respect to all of itself, but with respect to a particular   30
capacity and a particular part (just as [it is] capable of seeing and
capable of locomotion), which [part] is also manifest to sense perception. The parts of this sort, in fact, are in the female that
which is called the uterus, and in the male those associated with
the testicles or the male genitalia in all the blooded [animals]; for
while some of them have testicles, others have channels of that   35
sort. And there are differences between the female and the male   716b
in the bloodless [animals] as well, as many of them as have this
opposition. In the blooded [animals] the parts that serve for intercourse differ in their shapes. But one must understand that if
a small ruling beginning is altered, many of the [things that come]
after the ruling beginning usually change along with it. And this   5
[is] clear in the case of castrated [animals]; for when only the
generative part is destroyed, pretty much the whole shape
changes along with it to such an extent as to seem either to be
female or to fall little short [of that], which suggests that it is
with respect to no ordinary part and with respect to no ordinary
capacity that the animal is female or male. [It is] manifest, then,
that a certain ruling beginning comes to light, being the female   10
and the male. At all events, many [parts] change along with
[them] when [those parts] change through which [an animal is]
female or male, which suggests that a ruling beginning is undergoing a change.

Chapter 3

The [details] concerning the testicles and the uterus are not alike in all the blooded animals, and in the first place, those concerning
15   the testicles in males; for some of the animals of this sort don't have testicles at all, for example, the kind consisting of fish and that of snakes, but only two channels for seed; and others, while they have testicles, have them inside by the loin, in the region of the kidneys, and from each of the two a channel, as in those that
20   don't have testicles, [the channels] joining together into one as with those; [such are,] for example, among the [animals] that take in air and have a lung, all the birds and the egg-laying quadrupeds. For all these also have their testicles inside by the loin and two
25   channels from them like in snakes, for example, lizards and tortoises and all the horny-plated [animals]. The animal-bearing[7] [animals] all have their testicles in the front, but some of them [have them] inside at the end of the belly, for example, the dolphin, and not channels but a penis proceeding from these to the outside, as the ox-fish [do], while others [have them] outside, and of these
30   some [have them] hanging down, as a human being [does], while others [have them] by the anus, as pigs [do]. But distinctions have been made more precisely concerning them in our *Histories Concerning the Animals*.[8] The uterus has two parts[9] in all [the kinds of animals], just as the testicles in males are also two in [them] all.

---

7   The word translated as "animal-bearing" at 716b25 is *zōiotoka*, which is usually translated as "viviparous," i.e., "live-bearing," and which is contrasted with the word *ōiotoka*, meaning "oviparous," i.e., "egg-laying." But since fertilized eggs are themselves alive (cf. 731a5–7, 736a32–35), I prefer the term "animal-bearing" to "live-bearing."

8   Cf. *On the History of Animals* 509a27–510b5. The singular form of the Greek word translated as "histories" is *historia*," which means "inquiry" or "research" more primarily than it does "history" in today's sense. That primary meaning of the word, however, is reflected in the fact that we still speak of natural history, as in Aristotle's *On the History of Animals*, a work whose title, translated literally, is *On the Histories Concerning the Animals*.

9   The Greek term translated as "uterus," *husterai*, was understood as including not only the organ that is now called the uterus, but also the two oviducts that lead to it from the ovaries. In keeping with this understanding, the Greek word *husterai* is a plural form.

Some have this by the genitals, as do women, and [indeed] all those
[that are] animal-bearing not only externally but also within them-   35
selves, and as many fish as lay eggs that are seen externally, while   717a
others [have it] by the diaphragm, as do all birds and, among the
fish, those that are animal-bearing. The soft-shelled [animals] also
have a bifurcated uterus, and the cephalopods, for[10] the membranes that they have around their so-called eggs are uterine. It is   5
most undifferentiated in the octopuses, so as to seem to be single;
responsible[11] for this [is] the bulk of their body, which is similar
all around. And the [uterus] of insects is also bifurcated, in those
that are of [some] size; but in the smaller ones, it isn't clearly seen,
because of the smallness of their body. So the parts that have been   10
spoken about are [arranged] in animals in this manner.

Chapter 4

Concerning the difference among the organs for seed in males, if
someone is going to look into the causes on account of which they
are, [it is] necessary first to grasp what the structure of the testicles
is for the sake of. If, then, nature makes everything either on account   15
of the necessary or on account of the better, this part too would be
on account of one or the other of these. Now that it is not necessary
for coming into being [is] manifest; for it would belong to all the
[kinds] that generate, but as it is, neither snakes nor fish have testicles, yet they have been seen coupling and with their channels full   20
of milt. There is left then [that it is] for the sake of something better.

---

10 At 717a4, I read *ta gar*, with most of the manuscripts, instead of *kai ta* with
    Drossaart Lulofs, who follows manuscript Z. (Manuscript Z is the oldest, and
    generally regarded as the most authoritative among the manuscripts. But all
    editors agree that in some cases the readings of other manuscripts are more
    likely to be correct.) Drossaart Lulofs' text would be translated, instead of
    "for the membranes that they have...," as "and the membranes that they
    have...."

11 The Greek word that I have translated as "responsible" is *aition*, an adjectival
    form of the noun *aitia*, which means "cause." I might have chosen to translate
    the word as "a cause," or even as "the cause," but since Aristotle may perhaps
    have intended a substantive distinction between the adjective and the noun
    forms, I will continue to translate it as "responsible" or, if preceded by the
    definite article, as "what is responsible."

Now the work of most animals, just as [that] of plants, is pretty much nothing other than [the production of] seed and fruit. And just as in the [matters] concerning food those with straight intestines [are] more impetuous in their desire for food, so also those that don't have

25   testicles but only channels, or that have [them] but have them inside, [are] all quicker in the activity of their couplings. But those that need to be more moderate, just as in that sphere they don't have straight intestines, so also in this one their channels have convolutions, with a view to their desire being neither impetuous nor quick. And the

30   testicles are contrived for this; for they make the motion of the residue [consisting] of seed steadier, on the one hand, in the animal-bearing [animals], such as horses and the others of that sort and in human beings, preserving the doubling back [of the channels] (the manner of this [doubling back] ought to be studied from our *Histo-*

35   *ries Concerning the Animals*); for the testicles are no part of the channels, but they are attached—as women who are weaving attach

717b   stones to their looms—for when they are removed, the channels are drawn up inside, so that castrated [animals] are not able to generate, since if they hadn't been drawn up, they would have been able; and in fact a certain bull, copulating immediately after castration, impregnated [a cow], on account of its channels not yet having been

5   drawn up. On the other hand, in birds and the egg-laying quadrupeds [the testicles] receive the residue [consisting] of seed, so that its emission is slower than in fish. And [this is] manifest in the case of birds; for [in the times] around their copulations their testicles are much bigger, and in as many of the birds as copulate at one season, when this time has passed, their [testicles] are so small as to be

10   pretty much invisible, but around the [time of] copulation they are exceedingly big. Now those that have [their testicles] inside copulate more quickly; for those that have them outside don't emit their seed until they have drawn up their testicles.

Chapter 5

15   Further, while the quadrupeds have the organ for coupling, for it is possible for them to have it, it is not possible for birds and footless [animals], on account of the legs of the ones being under the middle of the belly and the others being wholly without legs,

whereas the nature of the penis is dependent on these and has its position there (wherefore also there comes to be the tension of the legs during intercourse; for the organ is sinewy and the nature of the legs is sinewy); so that since it is not possible [for them] to have this [organ], [it is] necessary that they also either not have testicles or not have them there; for to those that have them, both of them [i.e., testicles and penis,] have the same position. Further, for those that have their testicles outside, [it is] when the penis is heated through its movement [that] the seed is collected and comes forth, not as being ready immediately when [the mating animals] come in contact, as [it is] in fish. All the animal-bearing [animals] have their testicles in the front or outside except for the hedgehog.[12] For this alone has [them] by the loin, on account of the same cause as birds [do]. For [it is] necessary for their coupling to be quick; for they do not, like the other quadrupeds, mount upon the back, but they have intercourse [standing] upright, because of their spines. [The] cause, then, on account of which those that have testicles have [them] has been said, and [the] cause on account of which some have them outside and some inside.

20

25

30

Chapter 6

In as many [kinds of animals] as don't have [testicles], as has been said, [it is] not on account of the good, but of the necessary alone, that they don't have this part, and on account of its being necessary that their copulation be quick. And such is the nature of fish and that of snakes. For fish, on the one hand, copulate by coming alongside [the females] and they are quick to ejaculate. For just as [it is] necessary, in the case of human beings and all such [animals],

35

718a

---

12 At 717b26, Drossaart Lulofs would emend the Greek text, in keeping with a reading suggested by a 15th-century Latin translation, by adding words meaning "either inside" before those translated as "or outside," so that instead of "in the front or outside except for the hedgehog," the end of this sentence would be translated as "in the front, either inside or outside, except for the hedgehog." But it doesn't seem to me that the text needs to be emended, since "in the front" is easily understood as meaning "inside in the front," and it isn't necessary for Aristotle's argument to mention that external testicles are also situated in the front.

that they emit their semen while holding their breath, for those, this [emission] occurs [only] when they are not taking in sea water,
5 and they are easily destroyed when they aren't doing this; therefore, they need to not concoct the seed during their coupling, like the animal-bearing land animals, but they have their seed all together as a mass, concocted by the [heat of the] season, so that upon coming in contact with one another, they don't make [it],[13] but emit [it already] concocted. Wherefore they don't have testi-
10 cles, but channels that are straight and simple, as is a short portion [of the channels] near the testicles in quadrupeds; for while one part of the doubling back of the channel is bloody, the other is bloodless, which [part] receives,[14] and through which there proceeds, what is already seed, so that in these [animals] as well, when the semen arrives there, ejaculation comes about quickly. But in
15 fish, the entire channel is of such a sort as [it is] in the second part of its doubling back in human beings and such [animals].

Chapter 7

Snakes, on the other hand, copulate by winding themselves around one another, and they don't have testicles or a penis, as has been said earlier—[no] penis, because [they don't have] legs, and [no]
20 testicles, because of their length—but instead channels, like fish; for on account of their nature being elongated, if a further delay came about in the region of the testicles, the semen would become cold because of its slowness, which indeed happens in the case of those [animals] whose penis is big; for they are less fertile than those [with a] moderate-sized [one], on account of seed that is cold not being fertile, and [seed] that travels too far becoming cold. So
25 then the cause on account of which some animals have testicles,

---

13 At 718a8, I read *poiein*, with all the manuscripts, rather than *pettein*, with Drossaart Lulofs, whose reading is not supported by the manuscript tradition, but only by an early Arabic translation. According to Drossaart Lulofs' text, instead of "make [it]," the translation would read "concoct [it]."

14 At 718a13, Drossaart Lulofs would add the words *to hugron*, meaning "the liquid" after the word translated here as "receives." These words are not present in the manuscripts, and their inclusion is supported only by the Arabic translation referred to in n. 13, above.

while others don't have [them], has been stated. Snakes wrap themselves around one another because of their natural unfitness for
lying alongside [one another]. For since, though they are exceedingly long, they attach themselves with a small portion [of their
bodies], they don't fit together well; and so since they don't have
parts by which to grasp, instead of this they use the suppleness of   30
their body, winding themselves around one another. Wherefore also
they seem slower to ejaculate than fish, not only because of the
length of the channels, but because of the great care [that is taken]
in these [preliminaries].

Chapter 8

In regard to females, someone might be perplexed about the   35
[arrangements] having to do with the uterus, what their manner is;
for they contain many contrarieties. For neither are they the same
in all the animal-bearing [animals], but while human beings and all
the land animals [have the uterus] below by the genitals, the ani-   718b
mal-bearing selachians[15] [have it] above[16] by the diaphragm, nor
[are they all alike] in the egg-laying [animals], but fish [have the
uterus] below, just like human beings and the animal-bearing quadrupeds, but birds and as many of the quadrupeds as are egg-laying
[have it] above. Nevertheless, these contrarieties are actually in ac-   5
cord with reason. For, in the first place, the egg-laying [animals]
lay eggs in different ways; for some lay their eggs unperfected, for
example fish, for the [eggs] of fish are perfected and increase in size
externally. Responsible [for this is] that these [animals] are prolific,
and this [is] their work, just as in the case of plants; so if they perfected [their eggs] within themselves, [it would be] necessary [for   10
those] to be few in multitude; but as it is, they have so many that
each [of the two parts of the] uterus seems to be an egg, at least in

---

15 Selachians are a large class of cartilaginous fish that includes sharks, rays, and
    dogfish. According to Aristotle, all but one of these kinds of fish are animal-
    bearing (754a23–26).

16 When Aristotle uses directional terms such as "above" and "below" in an
    animal body, he is thinking of them in reference to the upright human body.
    Cf. *On the Progression of Animals* 706b3–10.

the very small fish; for these are the most prolific, as [is] also [the case] with the others whose nature is analogous to these, both among plants and among animals; for in these [what would have 15 been] increase in size is diverted to seed. But birds and the quadrupeds among the egg-layers lay perfected eggs, which have to be hard-skinned so as to be protected (they are soft-skinned as long as they are still increasing in size), and the shell comes into being by the agency of heat that evaporates the moisture out of the earthy 20 [matter]. [It is] necessary, then, that the region be hot in which this will happen. And such is the [region] around the diaphragm; indeed, this [region] concocts the nutriment. If, then, [it is] a necessity that the eggs be in the uterus, [it is] also a necessity that the uterus be by the diaphragm in those [animals] that lay their eggs [already] perfected, but [in those that lay them] unperfected, [that it be] below; for thus it will be on the way [toward their being laid]. 25 Also, it is more natural, where no other work of nature prevents it, for the uterus to be below than above; for its terminus is below; and where its terminus is, [there] also its work [is done]—and this [i.e., the uterus,] is where its work [is done].

Chapter 9

The animal-bearing [animals] also have a distinguishing feature in relation to one another. For some bring forth animals not only externally, but also within themselves, for example, human beings 30 as well as horses and dogs and all the [animals] that have hair, and among the aquatic [animals] dolphins as well as whales and the cetaceans of that sort.

Chapter 10

But while selachians and vipers bring forth animals externally, they first produce eggs within themselves. And they produce a perfected egg; for in this way the animal is generated out of the egg, whereas nothing [is generated] out of an unperfected [one]. [That] they 35 don't lay eggs externally [is] on account of their being cold in their nature and not, as some assert, hot.

14

Chapter 11

At all events, the eggs they generate [are] soft-skinned; for on account of their having little heat their nature doesn't dry the outermost [part]. So then on account of their being cold they generate
soft-skinned [eggs], and on account of [these] being soft-skinned,   719a
[they are not laid] externally; for they would be destroyed. Whenever an animal [of this sort] comes into being out of the egg, most
[things] come to pass in the same way as in small birds, and [the
eggs] descend below and become animals by the genitals, just as
in those that bring forth animals straightaway from the beginning.   5
Wherefore also the uterus in such [animals] is unlike both that in
the animal-bearers and that in the egg-layers, on account of its
having a share in both forms; for all the [animals] that are selachian-like have it both by the diaphragm and extending below.
(But both concerning this and concerning the other [kinds of]
uterus, one needs to have studied how they are from the *Dissections*   10
and the *Histories*.)[17] So that on account of their being egg-layers
of perfected eggs they have [the uterus] above, but on account of
bringing forth animals, [they have it] below, and they have a share
in both [forms]. But those [animal-bearers] that bring forth animals straightaway all [have the uterus] below; for no work of nature
prevents [it], nor do they generate in two stages. In addition to
these [things], [it is] impossible for animals to come into being by   15
the diaphragm; for [it is] necessary for the fetuses to have weight
and motion, but the place [by the diaphragm], since it is vital for
living, wouldn't be able to tolerate these. Furthermore, [there is]
a necessity that delivery [would] be difficult on account of the
length of the passage, since even now, in the case of women, if
around the time of childbirth they draw [the uterus] up by yawning or doing something of that sort, their delivery is difficult. And   20
even when the uterus is empty, if it is brought upwards, it causes
choking; for [there is] a necessity for the uterus that is going to
hold an animal to be stronger, wherefore all such [uteruses] are
fleshy, whereas those by the diaphragm are membranous. And even
in the case of those animals that generate in two stages, [it is] man-   25

17 Cf. *On the History of Animals* 510b5–511a34.

15

ifest that this happens; for they carry their eggs above and to the side, but the animals in the lower part of the uterus. The cause on account of which the [things] having to do with the uterus are contrary in some animals [to those in others], and in general why some
30   have it below and others above by the diaphragm, has been stated.

Chapter 12

As for why all [animals] have the uterus inside, whereas some have the testicles inside but others outside, responsible for the uterus being inside in all [is] that that which is coming into being, which needs protection and shelter and concoction, is in this, but the
35   outside of the body is easily injured and cold. And the testicles are
719b   outside in some but inside in others[18] on account of these too needing shelter and a covering, for preservation and for the concoction of the seed; for if they have been cooled and stiffened, [it is] impossible [for them] to be drawn up and emit the semen. Wherefore the testicles, in as many [males] as have [them] in the open, have
5   a shelter of skin, the so-called scrotum; but for all those the nature of whose skin is contrary, because of hardness, to its being suitable for enveloping and [to its being] soft and skin-like,[19] for example, those whose skin is fish-like and those [whose skin is] horny-plated, [it is] necessary for these to have [them] inside. Wherefore dol-
10   phins and as many of the cetacean-like [animals] as have testicles have them inside, as do, among the horny-plated, the egg-laying quadrupeds. And the skin of birds [is] hard, so that it is unfit for enveloping [the testicles] in [close] conformity with [their] size, and [so that] for all these animals there is this cause [of them being inside], in addition to those that have been mentioned earlier stem-
15   ming from the necessary circumstances of copulation.[20] And on

18 Drossaart Lulofs thinks that there is a lacuna in the manuscripts here, apparently because Aristotle doesn't give a reason why the testicles are outside in some kinds of animals. But he has already given one at 717a23–a36ff., a reason suggesting that this is in principle the best position for them.

19 At 719b7, Drossaart Lulofs would delete the words translated as "and skinlike," relying on no authority other than their apparent absence in the Greek text used as the basis for the Arabic translation referred to in nn. 13–14, above.

20 Cf. 717b14–23 (and 717b26–31).

account of the same cause, both the elephant and the hedgehog have their testicles inside; for their skin too is not naturally suitable for having the sheltering part separate. The uterus, in its position as well,[21] is situated in a contrary way in those [animals] that are animal-bearing within themselves as contrasted with those that are externally egg-laying, and among the latter, in those that have the   20
uterus below as contrasted with those [that have it] by the diaphragm, for example, in fish in relation to birds and to the egg-laying quadrupeds, and [contrary also] in those that generate in both ways, [first] producing eggs within themselves and [then] bringing forth animals into the open. Those that are animal-bearing both within themselves and externally, for example, human being and cow and dog and the others of that sort, have the uterus   25
near the belly; for it is advantageous for the preservation and growth of the fetuses that no weight press upon the uterus.

Chapter 13

In all these [animals], moreover, the channel through which the dry residue comes out is different from that through which the wet   30
[residue does]. Wherefore all the [animals] of this sort, both the males and the females, have genitals through which the wet residue is expelled and also, for the males, the seed, and for the females, the fetus.[22] This channel is above and in front of that for the dry

---

21 As it appears from the continuation of this sentence, Aristotle means by the "position" of the uterus either the forward, i.e., ventral, or the rear, i.e., dorsal, position, as distinct from its placement either "below" by the genitals or "above" by the diaphragm, which he had spoken of previously as an example of contrariety in "manner" (718a35–b1ff.). David Balme has suggested that this latter contrariety is appropriately not spoken of as one of position, since even in those animals in which the uterus (including the oviducts) is up by the diaphragm, it still ultimately leads to a conduit in the region of the genitals from which the egg or fetus is expelled, so that the difference between the two cases is not one of position but rather of prolongation and shape. *Aristotle De Partibus Animalium 1 and De Generatione Animalium 1*, translated with notes by D. M. Balme (Oxford: Oxford University Press, 1972/1992), 138.

22 At 719b33, the word translated as "fetus" is *kuēma*. In the two previous instances, however, where I have used the word "fetus" (719a15, b26), the Greek word was *embruon*. There as here, I have chosen the translation "fetus" in-

35 nutriment. But as many [animals] as lay an egg, but an unperfected
720a one, for example, as many fish as are egg-laying, these don't have
the uterus underneath the belly but by the loin; for the growth of
the egg doesn't prevent [this], on account of the growing [egg]
being perfected and developing externally.[23] And the channel is
the same as that for the dry nutriment in the [animals] that don't
5 have a generative genital organ, [namely,] all the egg-layers,[24] even
those of them that have a bladder, for instance tortoises; for the
channels are double for the sake of generation (*genesis*), not for the
excretion of the wet nutriment; and on account of the nature of
the seed being wet, the residue of the wet nutriment shares the
10 same channel. And this [is] clear from the fact that all animals produce seed, whereas not all have residue that is wet. Since, then,
the channels for seed in males, as well as the uterus in females,

---

stead of "embryo" because Aristotle is speaking of something that has con-
siderable weight, and this is not the case with what we now speak of as an embryo, which in humans typically weighs less than an ounce eight weeks after
conception, when the embryonic stage is said to end and the fetal stage to
begin. On the other hand, Aristotle sometimes also uses the term *kuēma* for
a newly developing animal at a stage much earlier than that at which we would
call it a fetus. As for the term *embruon*, in other authors (and at least once in
Aristotle) it refers to a young animal that has already been born, and in this
treatise Aristotle seems to use it mostly in reference to the fetal stage of pre-
natal development. But it isn't clear to me that this is how he uses it in all
cases. Since, then, I'm not certain how he understands the difference between
the two Greek terms, in subsequent passages I will translate each of them as
I think appropriate in keeping with modern usage, but where the Greek is a
form of *embruon* (which is used less often than *kuēma*), I will from here on
add the transliteration in parentheses.

23 Drossaart Lulofs brackets this sentence, along with the preceding one, and
he suggests in his critical apparatus that in his view they should either be
deleted or inserted at the end of the previous chapter. But they make sense
in this context, and as Balme notes (138), there is also a linguistic reason for
keeping them here—namely, that the Greek particle *oute* at 720a2 looks forward to *te* at a3.

24 Aristotle's assertion here that all the egg-layers, including tortoises in particular, lack a generative genital organ is in apparent contradiction with his earlier claim, at 717b14–15 (cf. b4–6), that the quadrupeds have the organ for
coupling. It is in keeping, however, with what he writes about the egg-laying
and blooded quadrupeds in *On the History of Animals* 503a4–6 (cf. 502b28ff.).
Presumably, then, in speaking of "the quadrupeds" in the earlier passage,
Aristotle meant quadrupeds in general, but not all of them.

need to be fastened and not wander about, and [it is] necessary for
this to happen either toward the front of the body or toward the
back, in the animal-bearers, because of the fetuses (*ta embrua*), the   15
uterus [is] in the front, while in the egg-layers [it is] by the loin
and the back.[25] But as many as bring forth animals externally after
having produced eggs within themselves, these have it [situated]
in both ways on account of their having a share in both, i.e., being
both animal-bearers and egg-layers; for the upper [parts] of the
uterus, i.e., where the eggs come into being, are under the dia-   20
phragm by the loin and the back, but as it continues downward,
[it is] in the belly; for there they are already bearing animals. And
in these too the channel for the dry residue and for copulation is
one [and the same]; for none of these has a penis, as has been said
earlier,[26] hanging down. Also, the channels of the males, both those   25
that have testicles and those that don't, are [situated] in the same
way as the uterus of the egg-layers; for in all [of them] they are attached at the back and along the region of the spine, for they need
to not wander about but to be stable, and of that sort is the rear-
ward region; for this provides continuity and fixedness. And so for   30
those that have their testicles inside, they are fastened immediately,
together with the channels,[27] and likewise for those that have them
outside[, the channels are fastened]; they [i.e., the channels] then
come together into one toward the region of the penis. And the

25 This statement, which presumably means that all the egg-layers have their
    uterus by the loin and the back, is in apparent contradiction with Aristotle's
    earlier claim, at 719b19–22, that those egg-layers that have their uterus below,
    such as fish, have it in the contrary position from those such as birds and the
    egg-laying quadrupeds that have it by the diaphragm; for since he says that
    the egg-laying fish have it by the back (719b34–720a2), one would have ex-
    pected him to say that the birds and the egg-laying quadrupeds have it in the
    contrary position, i.e., in the front. I am unable to resolve this difficulty.

26 Aristotle may be thinking of 720a5, where he has said that none of the egg-
    layers has a generative genital organ. Cf. n. 24, above.

27 Drossaart Lulofs would delete the phrase translated as "together with the
    channels," even though it appears in all the manuscripts—presumably be-
    cause he thinks that the "they" that "are fastened immediately" ought to be
    the spermatic channels, rather than the testicles, if the sentence is to make
    sense as a whole. But it seems to me that by interpreting the sentence in accord
    with the additions that I supply in brackets, the manuscript reading can be
    preserved.

channels are [situated] similarly in dolphins as well; but they have
35   their testicles hidden under the cavity surrounding the belly. And
so how [things] are positioned with regard to the parts that contribute to coming into being, and on account of what causes, has
720b   been stated.

Chapter 14

In the case of the other animals, those that are bloodless, the manner
5   of the parts that contribute to coming into being [is] not the same,
neither as in the blooded [animals] nor among themselves. There
are four kinds remaining, one, that of the soft-shelled, second, that
of the cephalopods, third, that of the insects, and fourth, that of
the shell-skinned (of these [last], while [it is] unclear concerning
[them] all, [it is] manifest that most of them do not couple; in what
manner they are constituted must be stated later). The soft-shelled
10   [animals] couple as do those that urinate backwards, when, the one
on its back and the other on its front, they interlock their tails; for
the tails, with their long appendage consisting of flaps, prevent
them from mounting front to back. The males have narrow seminal
channels, while the females [have] a membranous uterus, divided
15   along each side of the gut, in which the egg comes into being.

Chapter 15

The cephalopods intertwine near the mouth, pressing against one
another and spreading out their tentacles, and they intertwine in
this manner from necessity; for nature, by bending back the terminus of [the channel for] their residue, has brought it alongside the
20   mouth, as has been said earlier.[28] The female in each of these animals manifestly has the uterine part; for it holds an egg, undifferentiated at first, [but which] then, by separating apart, becomes
many, and she lays each of these unperfected, as do those of the fish

---

28 This isn't a reference to anything said earlier in this treatise, but apparently
to *On the Parts of Animals* 684b34ff. Some manuscripts even include at the
end of this sentence a phrase meaning "in the speeches about the parts."

that lay eggs. And the channel [is] the same for the residue as for   25
the uterine part, both in the soft-shelled [animals] and in these, for
it is where they emit their ink²⁹ through the channel [for it]. These
are on the underside of the body, where the mantle is extended and
the sea-water enters in; wherefore coupling of the male with the female comes about there; for [it is] necessary, if indeed the male emits   30
something, whether seed or a part or some other power, [for it] to
approach the uterine channel. But the male's putting his tentacle
through the [female's] funnel in the case of the octopuses, on which
grounds the fishermen assert that they copulate by means of a tentacle, is for the sake of intertwining, not because [it is] an organ useful for generation (*genesin*); for it is outside the channel and the body.   35
Sometimes also the cephalopods couple [by the one mounting] onto
the [other's] back; but it hasn't been seen yet whether [this is] for   **721a**
the sake of generation (*geneseōs*) or on account of some other cause.

Chapter 16

Of the insects, some couple, and their coming into being is from
animals that have the same name, as in the case of the blooded [ani-
mals], for example, locusts and cicadas and spiders and wasps and   5
ants; others couple and generate, but not [offspring] of the same
kind as themselves, but only larvae, nor do they come into being
out of animals, but out of rotting wet (pl.) [stuff], though some
out of dry (pl.), for example, fleas and flies and beetles; and others
neither come into being from animals nor do they couple, as for
instance gnats and mosquitoes and many such kinds. The females,   10
in most of those that couple, are bigger than the males. And the
males are not seen to have seminal channels. And to speak for the
most part, the male doesn't put any part into the female, but the
female [does] into the male, up from below. This has been observed   15

---

29 At 720b26, the word translated as "ink," *tholon*, is not present in the manuscripts, though it seems to have been present in the text used as the basis for
the Arabic translation referred to earlier. And this claim about where the cephalopods emit ink would be in keeping with what Aristotle says at *On the Parts
of Animals* 679a1–2, and *On the History of Animals* 524b20–21. What the
manuscripts have instead of *tholon* is the word *thoron*, or milt, which makes
no sense in a sentence about females.

in many cases, and likewise concerning the mounting,[30] while the opposite [has been observed] in a few; but [enough] hasn't yet been seen so as to make distinctions by kind[s]. This[31] is pretty much [the case] also in most egg-laying fish and in the egg-laying quadrupeds; for the females are bigger than the males on account of its

20   being advantageous with a view to the bulk that comes to be in them, because of their eggs, during pregnancy. In the females among them [i.e., the insects], the part analogous to the uterus, in which the embryos come into being, is split along the gut, as in the other [kinds] as well. And this [is] clear in locusts and in as many of them as, being of a nature so as to couple, are of [some]

25   size; for most of the insects are too small. So then the organs involved in coming into being for animals, about which it hasn't been spoken earlier, are [arranged] in this manner. And of the uniform [substances in the body, our speech] about semen and milk has been omitted, about which it is time to speak, about semen now

30   and about milk in what follows.

Chapter 17

Some animals manifestly emit seed, for instance as many of them as are blooded in their nature, but as for insects and cephalopods, in which of the two ways [they copulate is] unclear. So this must be looked into, whether all males emit seed or not all, and if not all, on

35   account of what cause some do while others don't; and also whether

721b   females contribute any seed or not, and if not seed, whether nothing else either, or [whether] they contribute something, but not seed. And further also, for those that emit seed, one must examine what they contribute, through their seed, to coming into being, and what

---

30 Drossaart Lulofs would delete that phrase translated as "and likewise concerning the mounting," even though it occurs in all the manuscripts.

31 It seems likely, and in keeping with the explanatory clause that follows ("for the females…"), that the referent of the word translated as "this" is the fact that the female is bigger than the male, not that the female puts some part of itself into the male. Accordingly, Balme has suggested (33) that the sentence at 721a11–12 ("The females, in most of those that couple, are bigger than the males.") ought to be transposed so as to immediately precede this one. He may be right.

the nature of seed is altogether and that of what are called the   5
menses, in as many of the animals as emit this fluid. Now it is
thought that all [animals] come into being out of seed, and the seed
out of those that generate. Wherefore it belongs to the same account
[to say] whether both the female and the male emit [it] or only one
of them, and whether it comes from all the body or not from all; for   10
it is reasonable,[32] if not from all, [that it] also not [come] from both
of those that generate. Wherefore one must examine in the first
place, since some assert that it comes from all the body, regarding
this, how it stands. Now there are pretty much four pieces of evidence that someone could use in support of the view that the seed   15
comes from each of the parts:[33] first, the intensity of the pleasure;
for when more of the same condition comes to be, [it is] more pleasant, and the condition [occurring] in all the parts is more than that
occurring in one or a few of them. Further, [the fact] that mutilated
[offspring] come into being from mutilated [parents]; for they assert
that because of their lacking the part, seed doesn't proceed from it,
and that the [part] from which [seed] doesn't come turns out not to   20
come into being. And in addition to these [pieces of evidence], the
likenesses [of offspring] to those that generated [them]; for they
come into being like [them], just as with respect to the body as a
whole, so also with respect to parts [as compared] to parts. If, then,
for the whole, [the fact] that the seed comes from the whole [is] responsible for the likeness, so also, for the parts, [the fact] that something comes from each of the parts would be responsible [for the
likenesses]. Further, it would seem to be reasonable [that], just as   25
there is something in the case of the whole out of which [the animal]
first comes into being, [there is] also [something] in the case of each
of the parts, so that if [there is] a seed for that, there would also be,
for each of the parts, some seed of its own. Plausible also [is] evidence of the following sort for these opinions; children come into
being resembling their parents not only in their inborn [character-   30

---

32 The Greek word translated here, and later in the chapter, as "reasonable"
    (*eulogon*) can also mean "plausible," though there is a different word, *pithanon*,
    appearing still later in the chapter, that means "plausible" unequivocally.

33 At 721b13–14, Drossaart Lulofs would delete the words translated as "in sup-
    port of the view that the seed comes from each of the parts," which words are
    indeed absent in some of the better manuscripts.

istics] but also in their acquired ones; for [in some cases,] when those
who generated [them] had scars, some of their offspring had the out-
line of the scar in the same places, and in Chalcedon, [in an instance]
where the father had a tattoo on his arm, the figure showed up as a
mark on his son, though jumbled together and not distinctly artic-
35   ulated. [It is] pretty much from these [things] above all, then, [that]
722a   some people are persuaded that the seed comes from all [the body].

Chapter 18

But [what] appears when [we] examine the argument closely [is]
the opposite, rather; for [it is] not hard to refute what has been
said, and in addition, it follows [from their position] that one
[must] say other, impossible, [things]. First, [to show] that likeness
is no indication of [the seed] coming from all [the body], [there is
5   the fact] that [offspring] come into being like [their parents] also
in voice and nails and hair and their motion, from which [things]
nothing comes. And [there are] some [characteristics] they don't
yet have when they are generating, for instance, a growth of gray
hair or of a beard. Further, [offspring] resemble ancestors from
further back, from whom nothing has come; for likenesses recur
after many generations, for example, the woman at Elis who had
10   intercourse with the Ethiopian; for not her daughter, but that one's
son, was born [looking] Ethiopian. And [there is] the same argument in the case of plants; for [it is] clear that in these as well the
seed would come into being from all the parts. But many [plants]
don't have some [parts], and someone could remove others, while
15   still others are adventitious growths. Further, [seed] doesn't come
from the seed-cases; and yet these too come into being having the
same shape. Further, does [the seed] come only from the uniform
[parts], from each [of them], such as flesh and bone and sinew, or
also from the non-uniform [ones], such as face and hand? For if
20   only from those . . . but these, the non-uniform [ones], such as face
and hands and feet,[34] resemble the parents more; so if [it is] not

---

34 Drossaart Lulofs would delete the words translated as "the non-uniform
[ones], such as face and hands and feet," even though they are found in all
the manuscripts. If they are genuine, as I see no reason to deny, they would

24

because of [the seed] coming from all [the body that] these are not [similar], what prevents those as well from being similar, not by [the seed] coming from all [the body], but on account of another cause? But if [the seed comes] only from the non-uniform [parts], then [it is] not from all [the parts]. And it is fitting [that it come] rather from those; for those are prior, and the non-uniform [parts] are put together out of those, and just as [offspring] come into being resembling [their parents] in face and hands, so also in flesh and nails. But if [it comes] from both, what would be the manner of its coming into being? For the non-uniform [parts] are composed out of the uniform [ones], so that to come from those would be to come from these and their combination; just as if something came from a written word, if [it came] from all [of it], it [would come] also from each of the syllables, and if from these, from the letters and their combination. So that if flesh and bones are constituted out of fire and the [bodies] of that sort [i.e., the elements], [the seed] would be from the elements rather;[35] for how is it possible from their combination? Yet surely, without this [combination], [the offspring] wouldn't be similar [to their parents]. And if something fashions this [combination] later, that [something] would be what is responsible for the likeness, but not the [seed's] coming from all [the body]. Further, if the parts are separated [from one another] in the seed, how are they alive? But if [they are] connected, they would be a small animal. And how [will] the [facts] having to do with the genitals [fit into this account]? For what comes from the male and from the female [will] not [be] alike. Further, if [the seed] comes from both [parents] equally, from all [the parts], two animals come into being; for each of the two will

---

presumably be meant to erase any doubt that the words translated as "those" and "these" in this sentence mean "the former" and "the latter." Also, the grammatical awkwardness in my translation of the beginning of this sentence is present also in the Greek.

35   At 722a35, I read *mallon*, meaning "rather," with Drossaart Lulofs, who follows the majority of the manuscripts. But there is a case to be made for preferring the reading of Z, which has the word *monon*, meaning "only," instead of *mallon*; for though that reading would be unexpected in the light of the preceding argument, it seems more consistent with the question that immediately follows.

have all [the parts].[36] Wherefore it also looks as if Empedocles, if one must speak in such a manner, says [things that are] most con-
10 sistent with this argument, [speaking finely] to that extent at any rate; but if [it is better to speak] in some different manner, [what he says is] not [spoken] finely.[37] For he asserts that there is a sort of tally in the male and [in] the female, but that a whole doesn't come from either of the two,

> But separated asunder is the nature of the limbs, one part in the man's . . . [38]

For [otherwise,] why don't the females generate out of themselves, if indeed [the seed] comes from all [the body] and they have a re-
15 ceptacle? But as seems likely, either [the seed] doesn't come from all [the body] or [it comes] in such a manner as that one says, [namely,] not the same [things] from each of the two [parents], which is why they need intercourse with one another. But even this [is] impossible. For just as when [the parts] are big, [it is] impossible for them to be preserved and to be ensouled if separated [asunder], as Empedocles generates [them] in the [period] of Love, saying,

20 > There many neckless heads sprang up,[39]

[and] then thus, he asserts, they used to grow together. But [it is] manifest that this [is] impossible; for neither would they, not hav-

---

36 At 722b7, I read *hekateron*, with the majority of the manuscripts, rather than *hekaterou* with Drossaart Lulofs, who proposed this emendation himself. According to his text, instead of "for each of the two will have all [the parts]," this clause would read "for [the seed] will have all [the parts] of each of the two." I'm not confident that the manuscripts are correct here, but I think that their reading makes tolerably good sense.

37 At 722b9–10, Drossaart Lulofs would delete the words translated as "to that extent at any rate; but if [it is better to speak] in some different manner, [what he says is] not [spoken] finely." But there is no ancient authority for doing so other than their apparent absence in the Greek text used as the basis for the Arabic translation referred to earlier.

38 Empedocles, fr. 63, in Hermann Diels, *Die Fragmente der Vorsokratiker*, rev. Walther Kranz, 6th ed. (Berlin: Weidmann, 1951–1952).

39 Empedocles, fr. 57 (Diels/Kranz).

ing soul or any sort of life, be able to be preserved, nor [would they], being like several animals, [be able] to grow together so as to be one again. But yet it turns out for those who assert that [the 25 seed] comes from all [the body] that they speak in this manner: just as [it was] then in the earth in the [period] of Love, so [it is] for these in the body. For [it is] impossible for the parts to come into being [already] connected and, coming together [with those from the other parent], to go off into one place. Next, how are the upper and lower and right and left and front and back [parts] separated asunder? All these [things] are unreasonable. Further, 30 the parts are distinguished, some by [their] capacity and others by [their] attributes, the non-uniform [ones] by being capable of doing something, for example tongue and hand, and the uniform [ones] by hardness and softness and the other such attributes; thus, not being in just any [condition are they] blood or flesh. [It is] clear, then, that [it is] impossible for that which has come away [from 35 the parts] to [rightly] have the same name as the parts, like blood from blood or flesh from flesh. But surely, if blood comes into 723a being out of something that is different, the coming away [of seed] from all the parts would also not be responsible for the likeness [between parents and offspring], as those who make such assertions say; for [it is] sufficient [for it] to come from one [part] alone, if indeed blood comes into being out of [what is] not blood. For 5 why couldn't all [the parts] come into being from one? This argument [of theirs] seems to be the same as that of Anaxagoras, that none of the uniform parts comes into being; except that that one [applied it] to all [things], whereas these make this out [to be the case] with regard to the coming into being of animals. Next, in what manner will these [parts] that have come from all [the body] 10 grow? For Anaxagoras asserts reasonably that flesh [coming] out of the nutriment is joined to the flesh; but for those who don't say these [things],[40] yet assert that [the seed] comes from all [the body], how, when [something] different is added, will it be bigger, unless that which joins with it changes? But yet if that which joins with it 15 is able to change, why is the seed not straight from the beginning of such a sort that blood and flesh are able to come into being out

---

40 Cf. Anaxagoras, fr. 11 (Diels/Kranz), "For in everything there is a portion of everything, except mind...."

of it, without it itself being both blood and flesh? For surely it isn't admissible to say this, that it grows bigger later on by means of the admixture, like wine when water has been poured in; for then each
20   [part] would have been most itself at first while it was [still] un-mixed; but as it is, it is more flesh and bone and each of the other [parts] later. Yet to assert that some [part] of the seed is sinew and bone, that statement is too far above us. In addition to these [things], if the female and the male become different [from one another] during gestation, as Empedocles says,

25   [The seeds] were poured into clean [vessels];
          some, on the one hand,
       become women, if they encounter cold…,[41]

but at all events both women and men manifestly change, just as [they become] fertile from infertile, so also [they become] male-en-gendering from female-engendering, which suggests that the cause is not in [the seed's] coming from all [the body] or not, but in that
30   which comes from the woman and from the man being commen-surate or incommensurate, or on account of some other such cause. [It is] clear then, if we shall posit in this manner, that [it is] not by [the seed's] coming from a certain [part that] the female [will come into being], so that not even the part that the male and the female [each] have peculiar [to itself will come into being because of this], if indeed the same seed is able to become both female and male,
35   which suggests that the part isn't in the seed. How is it different,
723b   then, to speak of this [part] or of the other parts? For if seed doesn't come into being from the uterus, there would be the same argument also with regard to the other parts. Further, some of the animals come into being neither from [animals] of the same kind nor [from ones] different in kind, for example, flies and the [various] kinds of
5   what are called fleas; and from these there do come into being ani-mals, but [animals that are] no longer like [these] in nature, but rather some kind of larvae. [It is] clear, then, that as many as are different in kind don't come into being from [a seed] that comes

---

41 Empedocles, fr. 65 (Diels/Kranz). The following clause reads a bit awkwardly, since there is no then-clause corresponding to the initial if-clause of the sentence.

from every part; for they would be like [their parents] if indeed likeness is a sign of [a seed's] coming from every [part]. Further, from one act of intercourse some even among the animals generate many [offspring] (in plants, this is altogether [the case]; for [it is] clear   10
that from one motion they bear all their fruit for the year); and yet how [would this be] possible if the seed were secreted from all [the body]? For [it is] necessary that one secretion come into being from one act of intercourse and [from] one separating off. And [it is] impossible that it be split apart in the uterus; for the splitting apart   15
would at that point be as it were from an animal, not a seed. Further, cuttings bear seed[, each of them] from itself; [it is] clear, then, that even before being cut off they bore the fruit from the same magnitude [within the plant], and that the seed didn't come from all the plant. But the greatest of these pieces of evidence [is one that] we have observed sufficiently in the case of insects. For even   20
if not in all [of them], at any rate in most the female extends some part of itself into the male during copulation; wherefore they also, as we said earlier,⁴² bring the copulation about in this way; for the ones below manifestly insert [something] into those above, not in all, yet in most of those that have been observed. So that it would   25
be manifest that not even in as many of the males as emit semen is its coming from all [the body] responsible for coming into being, but [it is caused] in some other manner, concerning which it must be examined later. For even if the [seed's] coming from all [the body] did occur, as they assert, it ought not to have been claimed that it comes from all [of its parts], but only from that which fashions, as from the carpenter but not from the material. But as it is,   30
[what] they say [is] the same as if [they had said that it came] even from the shoes; for a son resembling his father wears [shoes] that pretty much resemble [his]. And that a quite intense pleasure comes about during sexual intercourse, [it is] not the [seed's] coming from all [the body that is] responsible, but [the fact] that there is strong excitement; wherefore also if this intercourse happens often, the   35
enjoying [of it] becomes less for those who have [these] relations.   724a
Further, the enjoyment is near [to where] the completion [takes place], but it ought [on their view to be] in each of the parts, and not all at once, but in some earlier and in others later. And of mut-

---

42 Cf. 721a13–15.

ilated [offspring] coming into being from mutilated [parents],
[there is] the same cause as [the cause] why [offspring are] like their
5   parents. But non-mutilated [offspring] also come into being from
mutilated [parents], as do [offspring] unlike those who begat
[them]; concerning which [things] the cause must be looked into
later; for this problem is the same as those. Further, if the female
doesn't emit seed, [her not doing so is part] of the same argument
as [the seed's] not coming from all [the body]. And if it doesn't
come from all [the body], it is not at all unreasonable that it not
10   [come] from the female, but for the female to be responsible in some
other manner for coming into being. Concerning which it is to be
examined next, since [it is] manifest that the seed is not secreted
from all the parts. The beginning both of this inquiry and of those
15   that follow [is] to grasp first concerning seed, what it is; for thus
one will also be better able to contemplate concerning its operations
and the concomitants connected with it. Now seed in its nature
wishes to be the sort [of thing] out of which the [beings] constituted
in accord with nature first come into being, not because that which
produces [them] is something out of that, for example, [out of] the
20   human being; for [offspring] come into being out of this because
this[, that comes out of it first,] is the seed. And since one [thing]
comes into being out of another in many ways—in one manner, as
we assert that night comes into being out of day and a man out of
a boy, because this [comes] after this; and in another manner, as a
statue out of bronze and a bed out of wood and the other [things],
25   as many that come into being as we say come into being out of
material, the whole is out of something that is present in it and that
has been shaped; in a different manner, as unmusical [comes into
being] out of musical and as sick out of healthy, and in general as
the contrary out of its contrary. And further, besides these, as
Epicharmos portrays the escalation, out of the slander the abuse,
30   and out of this the battle; in all these, "out of something" [refers
to] the ruling beginning of the motion; and in some of such [cases]
the ruling beginning of their motion is within them, as in those just
spoken of (for the slander is a certain part of the whole disturbance),
while in others it is external, as the arts [are the ruling beginnings]
35   of the things produced and the lamp [is] of the burning house. As
for seed, [it is] manifest that it is in one of the two [classes]; for that
which comes into being is [out] of it either as out of material or as

30

out of that which first produced motion. For surely [it is] not as
this after this, for example, the sea voyage out of the Panathenaic
festival, nor as out of its contrary; for [it is] as it is being destroyed
[that] out of the contrary the [other] contrary comes into being,
and there needs [to be] something different underlying [them] out
of which, remaining present in [the new contrary], it will first be.
One must grasp, then, in which of the two [classes] seed is to be   5
placed, whether as material, i.e., acted upon, or as a sort of form,
i.e., acting [upon it], or indeed both. For at the same time, it will
perhaps be clear also how coming into being out of contraries belongs to all [beings] that [come into being] out of seed; for coming
into being out of contraries is indeed natural; for some [beings] do
come into being out of contraries, male and female, though others   10
out of one [parent] alone, for example, plants and also some of the
animals, in as many [kinds] as in which the male and the female are
not distinguished [as] separate. Now that which comes from the
generator, in as many [animals] as naturally couple, which first contains a ruling beginning of coming into being, is called semen,
whereas that which contains the ruling beginnings from both those   15
that have coupled (like the [seeds] of plants and of some animals,
in which the female and the male are not separated,) is called seed,
the first mixture, as it were, that comes to be out of female and
male, being like a sort of embryo or egg;[43] for these already have
that [which comes] out of both. Seed and fruit differ by the
earlier and later; for [it is] fruit by being out of another, whereas [it   20
is] seed by another [being] out of it, since in fact they are both
same. The primary nature of what is called seed[44] must again be
spoken of.[45] Now [it is] necessary that everything we find in the

---

43 At 724b18, I follow Drossaart Lulofs, who reads *ōion*, which I have translated
    as "egg," even though most of the manuscripts read *zōion*, meaning "animal."
    Cf. 731a5–6.

44 At 724b22, Z, with further support from the Arabic translation, would add the
    words *hōs gonēs*, or in translation, "as [meaning] semen," after the word trans-
    lated as "seed" here. And whether or not Aristotle actually wrote those words,
    the continuation of his account seems to indicate that the nature of semen is
    what he has in mind in speaking of "the primary nature of what is called seed."

45 Drossaart Lulofs would delete the three sentences beginning at 724b12 ("Now
    that which comes from the generator, . . .") and ending here, though without
    any manuscript authority for doing so.

body be either a part, [one] of those in accord with nature, and that
25 either among the non-uniform or the uniform ones, or [one] of
those contrary to nature, like a growth, or else a residue or a colliquament or nutriment. (I mean by "residue" what is left over from
the nutriment and by "colliquament" what is secreted out of the
[food ingested for] increase in size[46] by dissolution contrary to nature.) Now that it wouldn't be a part [is] manifest; for while it is uni-
30 form, nothing is composed out of it as out of sinew and flesh. And
further, it is not separate, whereas all the other parts [are]. Also, [it
is] not among the [parts] contrary to nature nor a deformity; for it
is present in all, and nature comes into being out of this. And nutriment [is] manifestly from outside. So that [it is] necessary for it
35 to be either a colliquament or a residue. Now it looks as if the ancients supposed it to be a colliquament; for to assert that it comes
from all [the body] because of the heat from the motion [of inter-
725a course] is equivalent to [calling it] a colliquament. But colliqua-
ments are a certain [class] of the [things] contrary to nature, and
out of the [things] contrary to nature nothing among the [things]
in accord with nature comes into being. [It is] necessary, then, for
it to be a residue. But now all residue is either from useless or useful
5 nutriment. By "useless" I mean [that] from which nothing further
is contributed toward nature, but if too much is consumed very
great harm is done, and by "useful" the contrary. Now [it is] manifest that it can't be the former sort of residue. For most of that sort
is present in those in the worst condition on account of age or sick-
10 ness, but [they have] the least seed; for either they don't have it at
all or [what they have is] not fertile on account of useless and morbid residue being mixed in. Seed, then, is a certain part of useful
residue.[47] And most useful is that which is last and out of which
each of the parts comes into being directly. For some [of the resi-

---

46 Regarding the distinction between nutriment and food ingested for increase
in size, see *On Soul*, 416b11–15.

47 As the sentence beginning at 725a14 ("Now residue from the first [stage of]
nutriment...") suggests, what Aristotle is referring to as "useful residue"
would seem to be, at least in part, residue from useful nutriment. Similarly,
what he will proceed to speak of as the "last" residue is presumably residue
from the last or final stage of nutriment. Regarding the stages of (the concoction of) nutriment, see, in the next chapter, 726b1–15 and also *On the
Parts of Animals* 650a2–35.

due] is earlier and some later. Now residue from the first [stage of] nutriment is phlegm and anything else that may be of that sort; for even phlegm is a residue from useful nutriment; a sign [of this is] that when mixed with pure nutriment it nourishes and is used up by those who are suffering from disease. But that which is last, from the greatest [amount of original] nutriment, comes to be smallest [in amount]. Yet one must consider that animals and plants increase in size from a small [increment] each day. For if a very small [amount] of the same [material] were added, their size would become excessive. One must say the opposite, then, of what the ancients said. For whereas they [said that seed is] that which comes from all [the body], we will say that seed [is] that which naturally goes to all [the parts], and whereas they [called it] a colliquament, it is manifestly a residue, rather. For [it is] more reasonable that what at the last goes to [all the parts] and what is left over from such [highly concocted nutriment] should be alike, as painters often have flesh-colored pigment left over [that is] like what has been used. But everything, when it decomposes,[48] is destroyed and departs from its nature. And a piece of evidence of [seed's] not being a colliquament but rather a residue [is] that the big animals have few offspring, whereas the small ones are prolific. For [it is] necessary that the big ones have more colliquament but less residue; for most of their nutriment is expended on their body, which is big, so that little residue comes into being. Further, no place is assigned in accord with nature for colliquament, but it flows wherever in the body it has free passage, but [there are places] for all the residues in accord with nature, for example, the lower stomach [for that] from the dry nutriment and the bladder [for that] from the wet [nutriment] and the upper stomach [for that] from the useful [nutriment] and the uterus and genitals and breasts for the [residues] that have the character of seed; for they collect and flow together into these [places]. And the concomitant [facts are] evidence that seed is what has been said; these happen because the nature of residue is of such a sort; for instance, weakness becomes evident after a very small amount of this has gone away, which suggests that the bodies are deprived

15

20

25

30

35

725b

5

---

48 At 725a27, the Greek participle translated as "when it decomposes," *suntēkomenon*, refers more particularly to dissolution or liquefaction, and is closely related to *suntēgma*, the word translated as "colliquament."

of the end [product] that comes from nutriment. (For some few,
10 during a short time in their young manhood, this gives relief by
going away, when there is too much, just as the first [stage of] nutriment does if it is excessive in amount; for also when this goes
away bodies are eased rather. Further, [there is relief] when other
residues go away along with [the seed]; for [it is] not only seed that
goes away in these cases, but other powers mixed in with it also go
15 away, and these are malignant—wherefore that which goes away
from some sometimes comes to be infertile on account of having
little of the capacity of seed. But for most and to speak generally,
[it is] rather weakness and lack of power [that] result from sexual
acts, on account of the cause that has been stated.) Further, seed is
20 not present either in the first stage of life or in old age or in periods
of illness, during illness on account of incapacity, in old age on account of nature not concocting [the nutriment] sufficiently, and
when [animals] are young on account of growth; for everything is
used up beforehand; indeed in five years, in the case of human be-
25 ings, at any rate, the body seems to acquire half of the total size that
there comes to be in the remaining time. In regard to these [things],
a difference shows up in many animals and plants, both in kinds,
in relation to [other] kinds and, within the same kind, in those of
the same form in relation to one another, for example, in a human
being in relation to a human being and in a vine in relation to a
vine. For some have much seed and others little seed, while others
30 [are] wholly without seed, not on account of weakness, but for some,
at any rate, on account of the opposite; for [their nutriment] is used
up for the[ir own] body, for example in some human beings; since
they are in good condition and are becoming very fleshy or rather
a little fat, they emit less seed and they have less desire for sex. And
35 similar [is] the condition in vines that "go goaty," which get out of
control because of their nutriment. ([They are said to "go goaty"]
726a since goats that are fat also copulate less, wherefore [their owners]
thin them down beforehand; and [people] speak of vines as "going
goaty" from the condition of the goats.) And [human beings] who
are fat, both women and men, appear to be less fertile than those
5 who aren't fat, on account of the residue that is concocted in the
well-fed becoming fat; for fat too is a residue, a healthy [one] on
account of good feeding. But some [living beings] don't even bear
seed at all, for example willow and poplar [trees]. Now there are

different causes of this condition too.[49] For because of lack of power
they don't concoct [their food], and because of power they use [it]
up in the way that has been said. Likewise also some have much   10
emission and much seed on account of power, whereas others [do]
on account of lack of power; for much useless residue is mixed together with [the seed], so that some even become ill when the purgation doesn't have free passage. And while some regain health,
others actually succumb; for [such residues] decompose there [i.e.,
into the emission], as also into the urine; for some have come down   15
with this ailment as well. Further, the same channel [serves] for the
residue and for the seed. And in those in whom there is residue
from both, from both the wet and the dry nutriment, the place
where the discharge of the wet [residue] occurs [is] where that of
the semen [does] (for it is a residue from wet; for the nutriment of
all [animals is] wet, rather), but in those that don't have this [dis-   20
charge of wet residue], [that of the semen occurs] where the solid
excrement is voided. Further, colliquescence [is] always malignant,
whereas the removal of residue [is] beneficial; but the voiding of
seed [is] of both [characters] on account of its taking along with it
[residue] from the non-useful nutriment. At any rate, if it were a
colliquescence, it would always cause harm, but as it is, it doesn't   25
do this.[50] So then that seed is a residue from useful nutriment and
from its last [stage of concoction], whether all [living beings] emit
seed or not, [is] manifest in what (pl.) has been said.

Chapter 19

After these [things] it must be determined what sort of nutriment
[seed] is a residue of as well as concerning menses; for menses   30
come into being in some of the animal-bearing [animals]. For
through these [things] it will be manifest also concerning the female, whether it emits seed like the male and what comes into

---

49 At 726a8, Drossaart Lulofs suspects that the text is corrupt. An alternative
    reading, though without any ancient support, would lead to a translation of
    this sentence as, "Now there are both [kinds of] causes of this condition too."

50 Drossaart Lulofs would delete the five and a half sentences beginning at
    726a11 ("for much useless residue is mixed together with [the seed], . . .") and
    ending here, though without any manuscript support.

being is a mixture out of two seeds, or whether no seed is secreted from the female, and if none, whether it also doesn't contribute
35   anything else toward coming into being but merely provides a
726b   place, or whether it contributes something, and [if] this, how and in what manner. Now that blood is the last [stage of] nutriment in the blooded [animals], and its analog in the bloodless [ones], has been said earlier.[51] And since also semen is a residue from nutriment and from the last [stage of nutriment], it would either be
5   blood or its analog or something out of these. And since [it is] out of blood as it is being concocted and divided up in some way [that] each of the parts comes into being, and [since] seed— though after being concocted it is discharged [with a character] quite different from blood—when it is unconcocted, and when someone forces it by frequently engaging in sex, has come out
10   bloody in some, [it is] manifest that seed would be a residue of the bloodlike nutriment that is distributed in its last [stage] to the parts. And because of this it has great power—for the removal of pure and healthy blood also tends to weaken—and the fact that offspring come into being resembling those who generated [them is] reasonable; for that which has gone to the parts [is] like what
15   is left over. And so the seed of the hand or of the face or of the whole animal is, without [their] determinacy, a hand or a face or a whole animal; that is, as each of those is actually, of such a sort is the seed potentially, either in virtue of its own bulk or [in that] it has some power within itself (for this [is] not yet clear to us
20   from the [things] that have been determined, whether the body of the seed is what is responsible for coming into being or whether it contains some disposition, or ruling beginning of motion, that is generative); for the hand too and any other of the parts, without a soul-ish[52] or some other power, is not a hand or any other part, but only the same in name. [It is] manifest also that, in as many

---

51 Cf. *On the Parts of Animals* 640a34–36, 651a14–15, 678a6–9.

52 At 726b22, the word translated here as "soul-ish" is *psukhikēs*, an adjective formed from the word for soul (*psukhē*) and a suffix that means "related to" or "fit for." It is an unusual word in Classical Greek, and I will continue to translate it as "soul-ish." (I use the form "soul-ish" rather than "soulish" primarily in order to avoid the New Testament connotations of the English word "soulish.")

[animals] as have a spermatic[53] colliquescence, this too is a residue.   25
This happens when it is dissolved into what came before [it], as
when the [coat] of plaster that has been laid on falls off at once;
for what comes away is the same as what was applied first. In the
same manner the last residue is the same as the first colliquament.[54]
And concerning these [things] let it have been determined in this   30
manner. But since [it is] necessary for more, and less concocted,
residue to come into being in the weaker [animal], and necessary
for it, being of such a sort, to be a large amount of bloodlike fluid,
and [since] the [animal] that according to nature has a lesser share
of heat is weaker, and [since] it has been said earlier that the female [is] of such a sort,[55] [it is] necessary also for the bloodlike   35
secretion that comes into being in the female to be a residue. And   727a
[it is] the discharge of what are called menses [that] comes into
being, [being] of such a sort. That the menses, then, are a residue
and that the menses in females [are] analogous to semen in males
[is] manifest. And signs that [this] has been said correctly are the   5
concomitants with regard to them. For at around the same age,
in males, semen begins to come into being and to be discharged
and, in females, the menses break forth, and their voices change
and the [changes] having to do with the breasts show themselves,
and as the prime of life comes to an end, the capacity for generating ceases in the ones and the menses in the others. Further,   10
[there are] also signs of the following sort that this discharge in
females is a residue. For the most part, women have neither
bleeding hemorrhoids nor a flow of blood from the nose nor any
other unless the menses come to a halt; and if any of these does
occur, the purgation becomes lesser, which suggests that the se-   15

---

53 At 726b25, the word translated as "spermatic" is *spermatikē*, which is closely
     related to *sperma*, the word that I have been translating as "seed." It generally
     means "related to seed" or "having the capacity of seed."

54 Drossaart Lulofs would delete the three sentences beginning at 726b24 ("[It
     is] manifest also that, . . .") and ending here, though he also suggests that they
     might belong after the passage that ends at 726a25, which passage, however,
     he also deletes (cf. n. 50, above). They do seem to belong at the end of that
     passage, but I leave them here out of deference to the manuscript tradition.

55 Aristotle has said this explicitly in *On Long-Livedness and Short-Livedness*
     466b14–16, though there is also a roughly similar claim in *On the Parts of
     Animals* 648a2–12.

cretion is diverted to these. Further, females have less prominent blood vessels and are less hairy and more smooth-skinned than males, because of the residue [that would go] to these being discharged along with the menses. And one ought to hold that this same [cause] is responsible also for the bulk of their bodies being

20 smaller in females than in males, among the animal-bearers; for in these alone does the flow of the menses come about externally, and of these, most noticeably in women; for a woman discharges the most [menstrual] secretion among the animals. Wherefore she is most noticeably always pale and without blood vessels that

25 show, and [there is] the manifest bodily deficiency [that] she has in comparison to males. Now since this is what comes to be in females like the semen in males, and [since] it isn't possible that two spermatic secretions both come to be [in the same being], [it is] manifest that the female does not contribute seed to coming into being. For if there were seed, there wouldn't be menses; but

30 as it is, because of these coming to be, there isn't that. It has been stated, then, why menses are a residue, as seed is. And someone might take as evidence for this some of the concomitant [facts] in animals. Those that are fat are less productive of seed than those without fat, as was said earlier[56] (responsible [for this is]

35 that fat too, like seed, is a residue, i.e., concocted blood, but not [concocted] in the same manner as seed. So that reasonably, since

727b the residue has been used up for fat, there is a deficiency of the [residues] having to do with semen), and among the bloodless [animals], cephalopods and the soft-shelled are at their best [for eating] during pregnancy; for on account of their being bloodless and their not coming to have fat, what is analogous to fat in them

5 is secreted into the spermatic residue. A sign that the female doesn't emit the sort of seed that the male does, and that [animals] don't come into being as a result of both [seeds] being mixed, as some assert, [is] that often the female conceives even though she hasn't had pleasure during intercourse; and on the contrary, when she does have it no less [than the male], and the

10 male and the female have kept the same pace,[57] [an offspring]

---

56 725b32–34.

57 At 727b10, I follow Drossaart Lulofs, who along with other modern editors would delete the word *para*, meaning "from," after the word translated here

doesn't come into being[58] unless the moisture of what are called menses is present [in a] suitable [amount]. For which reason, the female doesn't generate when [the menses] aren't coming to be at all, nor for the most part, if they are coming to be, while they are being exuded, but after the purgation. For at the former times, the power from the male that is present in the semen doesn't have nutriment or material out of which it will be able to constitute the animal, while at the other times, [the semen] is washed away along with [the menses] on account of their large quantity. But when, having come to be, they have gone away, what is left behind is constituted [into an animal]. And [as for] as many [females] as conceive when [menses] aren't coming to be, or while they are coming to be but not later, responsible for the former [is that] there comes to be as much moisture as is left behind after the purgation in those that are fertile, but there doesn't come to be any more residue so as to issue forth externally as well, and for the latter [it is that] the mouth of the uterus closes up after the purgation. Thus, when what has issued forth is a large amount, and the purgation, though it is still going on, is not so much as to cause the seed to be exuded along with it, if they have sexual relations then, they conceive. And [there is] nothing strange in [the menses] still coming to be in some who have conceived; for the menses recur, up to a point, even later, but in small quantity and not continually. But this is morbid, wherefore it occurs in few and infrequently; but what (pl.) comes to be for the most part is most in accord with nature. That the female, then, contributes the material for coming into being, and [that] this is in the composition that is the menses, and [that] menses are a residue, [is] clear.

15

20

25

30

---

as "have kept the same pace," even though there is no manuscript support for the deletion. The reading of the preponderance of the manuscripts doesn't seem quite grammatical, but if one were to accept it, the translation might read, instead of "the male and the female have kept the same pace," "they have kept the same pace, from the male also the female."

58 At 727b10, Drossaart Lulofs would read *outhen gennatai*, a reading supported only by the Arabic translation, instead of the manuscripts' *ou gignetai*, which I accept. According to Drossaart Lulofs' text, instead of "[an offspring] doesn't come into being," the translation would read "nothing is generated."

Chapter 20

As for what some suppose, that the female contributes seed during intercourse on account of there sometimes coming to be a delight in them nearly the same as that of males and, together with it, a wet
728a   discharge, this wetness isn't related to seed, but peculiar to the region in each of them. For it is a secretion from the uterus, and it comes to be in some but not in others; for it comes to be in those, to speak generally, who are pale-complexioned and feminine, but it doesn't come to be in those who are dark and masculine-looking.
5   And its amount, in those in whom it comes to be, is sometimes not in keeping with an emission of seed, but far exceeds it. Further, this or that food makes a great difference in there coming to be either less or more of such secretion, as for example, some of the bitter
10   [foods] make the discharge conspicuously large in amount. And that pleasure occurs during intercourse is not only from the emitting of seed, but also of *pneuma*,[59] out of which, by its coagulating, seed is derived.[60] [This is] clear in the case of boys who are not yet able to ejaculate, but are near the age for it, and in that of sterile men; for in all these there comes to be pleasure as they are being rubbed. And in those who are mutilated with respect to the coming into being [of
15   offspring], the bowels are sometimes loosened because residue that is not able to be concocted and to become seed is secreted into the bowel.[61] A boy resembles a woman in appearance also, and the woman is like a sterile male; for the female is [female] in virtue of a certain incapacity, namely, not being able to concoct seed from the
20   last [stage of] nutriment (this is either blood or its analog in the

---

59   Aristotle calls *pneuma* "hot air" at 736a1. It is sometimes hardly more than a synonym for *aēr*, meaning "air" (cf. 735b23–27; *On Sense Perception and Sense-Perceptibles* 443b4–5), but it is also the word for wind as well as for breath (as I translated it at 718a3). Because of the range of meanings of this word, and in particular because of the difficulty of assigning it either to non-living or to living nature, I will either not attempt to translate it, but merely transliterate it, as I do here, or else include a transliteration in parentheses.

60   Cf. 735b7–736a9.

61   This sentence perhaps belongs after the following one, even though it appears here in the manuscripts. Balme in his translation transposes it to the later position (49).

bloodless [animals]) on account of the coldness of her nature. So then just as in the bowels there comes to be diarrhea on account of lack of concoction, so in the blood vessels there are other hemorrhages as well as those of the menses; for this too is a hemorrhage, but those are on account of sickness, while this is natural. And so [it is] manifest that, reasonably, coming into being comes about out of this. For the menses are a seed that isn't pure but needs working on, just as in the coming into being connected with crops, the nutriment, when it has not yet been sifted, is in [the plant], but needs working on for purification. Wherefore when the former is mixed with semen, and the latter with pure nutriment, the one generates and the other nourishes. A sign of the female not discharging seed is also that the pleasure during intercourse comes to be from touch in the same region as in males; and yet they don't discharge this moisture from there. Further, this secretion doesn't come to be in all females, but in those with blood, and not in all these, but in those whose uterus is not by the diaphragm and that do not lay eggs, and further, not in those that don't have blood but [rather] its analog; for what blood [is] in some,[62] a different compound is in these. [The] cause of there not being a purgation, either in these or in the ones mentioned of those that have blood,[63] [is] the dryness of their bodies, which leaves little residue, that is, only so much as is sufficient for coming into being, but not for being emitted externally. And as for as many as are animal-bearers without [first] producing an egg (these are the human being and as many of the quadrupeds as bend their hind legs inward; for all these bring forth animals without producing an egg), [menses] come to be in all these, unless they are mutilated in their coming into being, for example the mule, but the purgations are certainly not abundant as in human beings. As for how these [things] occur with regard to each of the animals, it has been written

25

30

35

728b

5

10

---

62 At 728b2, I read *eniois*, with all the manuscripts, instead of *ekeinois*, a modern emendation first proposed by Platt and accepted by Drossaart Lulofs. If that emendation were accepted, the translation would read "those" instead of "some."

63 Here at 728b4, I follow Drossaart Lulofs in rejecting the words *tois katō ekhousi kai mē ōiotokousin*, which are found in some, but not all, of the manuscripts. This phrase would be translated as "[i.e.,] those having [the uterus] below and not laying eggs," thus making a claim that contradicts both what Aristotle has said in the previous sentence and what he will say in the next.

with precision in the *Histories Concerning the Animals*.[64] The greatest
15   purgation, among the animals, comes to be in women and the greatest discharge of seed in men, in proportion to size. Responsible [for
this is] the constitution of their body, which is wet and hot; for [it
is] necessary that the largest amount of residue come to be in such
[a body]. Further, they also don't have in their body the sort of parts
20   to which the residue is diverted, as [there are] in the others; for they
have neither a large amount of hair on their body nor secretions that
consist of bones and horns and teeth. A sign that the seed is in the
menses: as has been said earlier,[65] this residue comes to be in males
and there are indications of the menses in females at the same time
of life, which suggests also that the regions receptive of each of the
25   two residues are differentiated at the same time; and as each of the
two nearby regions becomes less dense, there bursts forth youthful
prime's growth of hair. When they are about to be differentiated,
the regions swell up as a result of *pneuma*, in males more noticeably
around the testicles, though there are indications also around the
30   breasts, and in females more around the breasts; for when they are
raised up about two fingers' breadth, then the menses come to be in
most of them. Now in all those, of the [beings] that have life, in
which the male and the female are not separate, the seed in these is
like an embryo. I mean by "embryo" the first mixture out of female
35   and male, wherefore also out of one seed one body comes into being,
for example, out of one grain of wheat one stalk, just as out of one
729a   egg [there comes into being] one animal (for double [yolk] eggs are
two eggs). But in all the kinds in which the female and the male are
distinct, in these it is possible for many animals to come into being
out of one seed, which suggests that seed in the plants differs by
5   nature from [that] in animals. A sign: from one act of copulation
more than one [offspring] comes into being in those that are able to
generate more than one. From which [it is] also clear that semen
doesn't come from every [part]; for separated [parts] wouldn't have
been secreted immediately from the same part nor would they, after
coming into the uterus together, have been separated there; but it
10   happens as [is] reasonable, since the male provides the form as well
as the ruling beginning of motion, whereas the female [provides]

---

64   *On the History of Animals* 572b29–573a17.

65   727a4–8.

the body, i.e., the material: as in the curdling of milk, the milk is the body, but the fig juice or the rennet is what contains the ruling beginning that sets it, so what is from the male, being divided up within the female, [contains the ruling beginning]. As for the cause from which in one case it is divided into a larger number, in another into a smaller one, and in another [it remains] single, [that] will be a different account. But because of its not differing at all in form, so long as what is divided is proportionate to the material, i.e., not too little, so as neither to concoct nor constitute [it], nor too much, so as to dry it out, in this way more [offspring than one] are generated. But out of what is [doing the] constituting primarily [i.e., proximately], out of [what is] now one, only one [offspring] comes into being. That the female, then, does not contribute semen to coming into being, but [that] it contributes something, and this is the composition that is the menses, and its analog in bloodless [animals], [is] clear from what (pl.) has been said as well as to people who examine in accord with reason generally. For [it is] necessary that there be that which generates and [that] out of which, and, even if this is one, that it differ at any rate in form and by the [defining] account [of each] of them being different, and in those that have the capacities separate, [it is necessary] also that both the bodies and the nature of that which acts and of that which is acted upon be different. If, then, the male is [male] as setting in motion and acting, and the female [is], insofar as it is female,[66] as being capable of being acted upon, the female would contribute to the semen of the male not semen but material, which [is] also manifestly [what] happens; for the nature of the menses is in accord with the primary [i.e., proximate,] material.

Chapter 21

Concerning these [things] let it have been determined in this manner. And at the same time [it is] manifest from these [what needs to be said] concerning the [questions] to be examined next, how   **729b**

---

66 At 729a29, Drossaart Lulofs would delete the word *hēi thēlu*, which I have translated "insofar as it is female," although they are present in the great preponderance of the surviving manuscripts. If I had omitted those words, my translation would also have changed the preceding "[is]" to "[is female]."

ever does the male contribute to coming into being and how is the
seed from the male responsible for that which comes into being,
whether as being present in and being immediately a part of the
body that is coming into being, mixing in with the material from
5   the female, or [is it that] the body of the seed doesn't participate
at all, but [rather] the capacity and motion within it; for this is
what acts, while that which is being constituted and acquiring its
shape [is] what remains within the female of the [menstrual] residue. According to reason, indeed, it appears so, as well as on the
basis of the facts. For it is manifest, to people who examine gener-
10   ally, that not one [thing] comes into being out of that which is acted
upon and that which acts, in the sense that that which acts is present in that which is coming into being, nor indeed in general out
of that which sets in motion and [that which] is set in motion. But
in truth the female as female [is] capable of being acted upon, and
the male as male [is] capable of acting and [is] from where the ruling beginning of the motion [is]. So that if the extremes of each
15   of the two are taken, insofar as the one [is] capable of acting and a
mover and the other capable of being acted upon and [something]
moved, the one [being] that comes into being is not out of these,
except in the sense that the bed [is] out of the carpenter and wood
or the sphere out of the wax and the form. [It is] clear then that
neither [is it] a necessity that something come away from the male
20   nor, if something does come away, is that which is generated, on
this account, out of it as [out of something] present in [it], but as
out of [something] that has set [the material] in motion and the
form, just as the one who has been healed is [healthy] from the art
of healing. And in agreement with reason [is] what (pl.) happens
regarding the facts. For on account of this some of the males, even
though they couple with females, manifestly don't put any part
25   into the female, but on the contrary, the female [does] into the
male, as happens with some of the insects. For what the seed ac-
complishes within the female for those who put [something] in,
the heat and capacity in the animal itself accomplish for these,
when the female brings into [it] the part that is receptive of the
residue. And on account of this, these sorts of animals are en-
30   twined together for a long time, but when they have separated they
generate quickly. For they stay coupled until they have done the
constituting, just as semen [does]; but when they have separated,

the embryo is released quickly; for [what] they generate [is] imperfect; for all such [animals] bring forth larvae. But what happens
with regard to birds and the class of egg-laying fish [is] the biggest
sign that neither does the seed come from all the parts nor does   35
the male put into [the female] any part of such a sort as will be   730a
present in the generated [being], but produces an animal only by
the capacity in the semen, just as we said in the case of the insects
in which the female puts [a part] into the male. For if the hen happens to be carrying wind eggs,[67] [and] if afterwards it is mounted   5
while the egg hasn't yet changed from being wholly yellow to becoming white, they become fertile instead of wind eggs, and if [it
is] mounted by a different [cock] while [the egg] is still yellow, the
brood of chicks, all of it, turns out in conformity with the one that
mounted later. Wherefore some of those who are serious about   10
highbred birds act in this manner, changing [the cocks between]
the first matings and the later ones, which suggests that [the seed]
isn't mixed in and present in [the egg] and that it doesn't come
from all [the body]; for it would have come from both [cocks], so
that [the offspring] would have had the same parts twice. But by
its capacity the seed of the male renders the material and nutri-   15
ment in the female of a certain sort. That which entered later is
able to do this by heating and concocting; for the egg takes nutriment as long as it is growing bigger. And the same [thing] happens
also in regard to the coming into being of egg-laying fish. For when
the female has laid its eggs, the male sprays milt over [them].[68]   20
And the ones that it touches, these become fertile, while the ones
that it doesn't [are] infertile, which suggests that the male doesn't
contribute, for the animals, to their [being] so big but to their
[being] of a certain sort. So then that the seed doesn't come from
all [the body], in those of the animals that emit seed, and [that]   25
the female doesn't contribute to the coming into being of the [off-

---

67 At 749a35–b1, Aristotle says that what some call wind eggs are eggs, or em-
    bryos as he says there, that arise spontaneously (i.e., without previous copu-
    lation) in female birds. Cf. 730a30–32.

68 Aristotle treats the male fish sprinkling eggs with his milt as being the same
    sort of activity as a cock mounting a previously mounted hen, because on his
    view most fish eggs, like most bird eggs, are not produced spontaneously but
    are the product of a previous act of copulation (756a15–18, a27–29ff.).

spring] that are being constituted in the same way as the male, but the male [contributes] the ruling beginning of motion and the female the material, [is] clear from what has been said. For on account of this, as [on the one hand] the female doesn't generate, itself by itself; for it needs a ruling beginning and what will set in
30   motion and give determinateness (but in some of the animals, for example birds, nature [in the female] is able to generate up to a point; for these [females] do constitute, but [what] they constitute [are] imperfect, the so-called wind eggs),

Chapter 22

so [on the other hand][69] the coming into being of the [offspring] that come into being occurs in the female; neither the male itself nor the
35   female emits semen into the male, but both contribute what comes
730b   to be from them into the female, on account of there being in the female the material out of which what is being fashioned is [constituted]. And [it is] necessary that some of the material, out of which the embryo is constituted in the first place, be present immediately in a mass and that some come in continually so that the gestating
5   [embryo] may grow. And so [it is] a necessity that birth take place in the female; for the carpenter also is close by the wood and the potter by the clay, and in general every act of making and final motion [is] by the material, as for example, housebuilding in the houses being built. And someone might grasp from these [things] also how the
10   male contributes to coming into being; for not every male emits seed, and in as many as do emit [it], this is no part of the embryo that is coming into being, just as also nothing goes from the carpenter as an addition to his material, the pieces of wood, nor is there any part of the art of carpentry in what is coming into being, but
15   from him, through the motion in the material, the shape and the form come to be in [it], and his soul, in which [there is] the form, and his knowledge move his hands or some different part with a par-

---

69 Here it seems especially worth recalling, as I said in the Introduction, that the chapter divisions are a relatively late addition to the textual tradition. I certainly do not understand why a division was made in the middle of this sentence.

ticular sort of motion, [the motions being of a] different [sort] from which that which comes into being is different, and [of] the same [sort] from which it is the same, and his hands [move] his tools, and his tools the material. Similarly also nature, that in the male of those [males] that emit seed, uses the seed as a tool and as having motions[70] in actuality, just as, in the case of the [products] that come into being in accord with art, the tools are set in motion; for the motion of the art [is] in a way in those. So then as many as emit seed contribute to coming into being in this manner; but as many as don't emit [seed], but [where] the female inserts one of its own parts into the male, seem to be doing the same as if someone had brought the material to the craftsman. For on account of the weakness of such males, nature [in them] is unable to act through [intermediaries that are] different, but the motions are barely strong enough with it sitting in attendance itself, and it resembles [clay] modelers, not carpenters; for it fashions what is being constituted, not being in contact [with it] through [something] different, but itself with the parts of itself.

Chapter 23

Now in all the animals that are capable of locomotion the female is separated from the male, and one animal is female and a different one male, but [they are] the same in form, for example, both [are] human being or horse. But in plants these capacities [are] mixed together, and the female is not separated from the male. Wherefore also they generate [by] themselves out of themselves, and they don't emit semen, but an embryo, what are called their seeds. And Empedocles says this finely, composing [the verse],

Thus do tall trees bring forth eggs; first olive trees....[71]

---

70 At 730b21, I read *kinēseis*, with the preponderance of the manuscripts, rather than *kinēsin*, an emendation accepted by Drossaart Lulofs and other modern editors, but supported only by a 15th-century Latin translation. According to that reading, instead of "motions in actuality" the translation would be "motion in actuality."

71 Empedocles, fr. 79 (Diels/Kranz).

For the egg is an embryo, and out of a certain [part] of it the animal comes into being, while the remainder [is] nutriment; and out of a part of the seed the [plant] that is springing up comes into being, while the remainder becomes nutriment for the shoot and for the
10   first root. And in a certain manner the same [things] occur also in the animals that have the female and the male separate. For whenever there is a need [for them] to generate, they become unseparated just as in plants, and their nature wishes to become one; which becomes apparent to sight when they join together and couple, [namely,] that a single animal comes into being out of both.[72]
15   And those that don't emit seed are of a nature to be entwined together for a long time, until they have constituted the embryo, for example, those of the insects that couple; but others, [only] until [the male] has discharged, from the inserted parts of itself, something that will constitute the embryo over the course of more time, for example in the case of the blooded [animals]. For while the for-
20   mer keep together for some part of a day, semen does its constituting over the course of several days, but after they have emitted that which is of such a sort, [those that have done so] separate. And animals seem to be just like plants split apart, as if in the case of those too, when they were bearing seed, someone were to break [them] up and separate [them] into the female and male that is present in [them]. And nature fashions all these [things] reason-
25   ably. For no work nor any action belongs to the being of plants other than the generation (*genesis*) of seed, so that since this comes about through the female and the male coupling, [nature] joined and arranged these together with each other; wherefore in plants the female and the male [are] unseparated. Concerning these, how-
30   ever, it has been examined in other [speeches], but generating [is] not the only work of the animal (for this is common to all living beings), but they all share in a sort of knowing, some more, some less, and some extremely little. For they have sense perception, and sense perception is a sort of knowing. As for the preciousness or worthlessness of this, it makes a lot of difference whether [we]
35   consider it in relation to intelligence or in relation to the class of

---

72 Drossaart Lulofs, with other modern scholars, would delete the phrase translated as "[namely,] that a single animal comes into being out of both," though
without any manuscript authority.

soulless [beings]. For compared to being intelligent, sharing in
touch and taste alone seems to be as if nothing, but compared to a
plant or a stone [it is] marvelous; for it would seem enough to be
content with to obtain even this knowing and not to lie dead and
not [even] being. [It is] by sense perception [that] animals differ
from the [beings] that are merely alive. But since [it is] a necessity
[for a being] also to be alive, if [it] is an animal, whenever there is
a need to accomplish the work of what is alive, then they couple
and join together and become as if a plant, as we have said. But
the shell-skinned among the animals, being in between animals
and plants, as they are in both classes, do the work of neither; for
as plant, [this class] doesn't have the female and the male and [a
member of this class] doesn't generate into another, but as an animal, it doesn't bear fruit out of itself like the plants, but it is constituted and generated out of a certain earthy and wet compound.
But concerning the coming into being of these it must be spoken
later.

Book Two

Chapter 1

The female and the male, that they are ruling beginnings of coming into being has been said earlier, and what their capacity [is]
and the articulation (*logos*) of their being.[1] But why the one comes   20
into being, and is, female and the other male, insofar as [it is]
from necessity and [from] the first mover, and [from] what sort
of material, the argument has to try to indicate as it proceeds,
but insofar as [it is] on account of the better and the cause for
the sake of something, it has its ruling beginning from higher
up. For since some of the [things] that are, are eternal and divine,
whereas the others admit of both being and not being, and [since]   25
the beautiful and the divine [is] responsible always, in accord
with its own nature, for what is better in the [things] that admit
[of it], and [since] what is not eternal admits of being[2] and of
sharing in both the worse and the better, and [since] soul is better
than body, and the ensouled than the soulless (on account of the
soul), and being than not being, and living than not living, on   30
account of these causes there is coming into being of animals;
for since the nature of such a class [is] unable to be eternal, in
accord with the manner that it does admit of, in accord with this
what comes into being is eternal. Now in number [it is] impossible—for the being of the [things] that are [is] in the particular;
and if it [i.e., what comes into being] were of such a sort [i.e.,   35
eternal in number or as a particular], it would be eternal—but

---

1   716a5–7, a13–23.

2   At 731b27, Drossaart Lulofs would add, after the words translated as "admits
of being," the words *kai mē einai*, which mean "and of not being." These
words are not present in any of the manuscripts, though they were apparently
in the text used as the basis for the Arabic translation.

732a   in species³ it is possible. Wherefore there is always a class of
human beings and of animals and of plants. And since a ruling
beginning of these [is] the female and the male, there would be
the female and the male in the [things] that are.⁴ And with the
cause, the one that first sets in motion—to which the articulation
5   (*logos*) and the form belong—being better and more divine in its
nature than the material, it is better also for the superior to be
separated from the inferior. On account of this, in as many [be-
ings] as admit of it and to the extent that it is possible, the male
is separated from the female; for the ruling beginning of motion,
which is present [as] the male⁵ in the beings that come into being,
is better and more divine, while the female is material. But the
10   male comes together and joins with the female for the work of
generation (*geneseōs*); for this is common to both. Now in virtue
of having a share in the female and the male [a being] is alive
(wherefore the plants too share in life), but in virtue of sense perception there is the class of the animals. And of these in pretty
much all those capable of locomotion the female and the male
15   are separate on account of the causes that have been stated. And
while some of these, as was said, emit seed, some do not emit
[any] in their coupling. And responsible for this [is] that the more
precious ones are also more self-sufficient in their nature, so that
they have a [greater] share in size. And this [is] not without soul-
ish heat; for [it is] a necessity that the bigger be set in motion by
20   a greater capability, and what is hot [is] capable of setting in mo-

---

3   At 731b35, the word I have translated as "in species" is the dative case of the
word *eidos*, which I have been translating as "form." This word means both
the form or character of a being and the class or species constituted by its
members resembling one another in some respect and thus being character-
ized by the same form. I will generally continue to translate *eidos* as "form,"
but if it seems more appropriate to translate it as "species," I will do so.

4   At 732a3, Drossaart Lulofs would change the phrase *en tois ousin*, which I
have translated as "in the [things] that are," to *en tois ekhousin*, which means
"in the [things] that have them," though without any manuscript support.

5   At 732a8, Drossaart Lulofs reads, instead of *hē arren huparkhei*, which I have
translated as "which is present [as] male," *hēi to arren huparkhei*, which would
be translated as "to which the male belongs." His reading finds support in
one rather late manuscript (which lacks, however, the definite article *to*) as
well as in a Latin translation by William of Moerbeke, but the one I have chosen is the reading of the preponderance of the manuscripts.

tion. Wherefore, looking over the whole, [it is possible] to say that the blooded [animals] are bigger than the bloodless and that those capable of locomotion [are bigger] than the stationary animals; which very ones emit seed on account of their heat and their size. And concerning male and female, [the] cause on account of which each of them is, has been said. Of the animals, some perfect and re-   25
lease externally [in each instance, an offspring] like itself, for example, as many as visibly bring forth animals, whereas others bring forth [a being that is] unarticulated and that hasn't taken on its own shape. And of those of this [latter] sort, the blooded [ones] lay eggs, while the bloodless [ones] bring forth larvae. And an egg differs from a larva; for an egg is [something] out of which the [animal] that   30
comes into being comes into being out of a part (while the remainder is food for the [animal] that comes into being), whereas a larva is [something] out of [the] whole of which [the] whole of the [animal] that comes into being comes into being. Now of those that manifestly bring to perfection an animal like [themselves] and that are animal-bearers, some produce animals straightaway inside themselves, for example, human being and horse and cow and, of those in the sea, dolphin and the others of that sort, whereas some, after first having   35
produced eggs inside themselves, bring forth animals externally, for   732b
example, those that are called selachians. And of those that lay eggs, some release the egg [already] perfected, for example, birds and as many quadrupeds as lay eggs and as many [animals as are] footless, for example, lizards and tortoises and the larger class of the snakes (for the eggs of these, when they have come out, no longer increase   5
in size), whereas some [release the eggs while still] unperfected, for example, the fish and the soft-shelled [animals] and those that are called cephalopods; for the eggs of these increase in size after coming out. All the animal-bearers[6] are blooded, and the blooded, as many as are not wholly sterile, either bring forth animals or lay eggs. Of the bloodless, the insects bring forth larvae, as many [of them]   10

---

6   At 732b8, Drossaart Lulofs includes, after the word translated as "animal-bearers," the words *ē ōiotokounta*, which mean "or egg-layers." These words do indeed appear in all the manuscripts. But Aristotle knows that not all egg-laying animals are blooded, as we see even from the previous sentence, which speaks of the egg-laying soft-shelled animals and cephalopods (cf. 715a28–b2) that lay unperfected eggs. And the words seem not to have been present in the text used as the basis for the Arabic translation, which doesn't translate them.

as either come into being from coupling or themselves couple.
For there are, among the insects, some of that sort, which come
into being spontaneously but are female and male, and something comes into being out of them when they couple, yet what
comes into being [is] unperfected; the cause has been stated ear-
15   lier in other [speeches].[7] There occurs much overlapping among
the kinds; for neither do all the bipeds bring forth animals (for
the birds lay eggs) nor do they all lay eggs (for the human being
brings forth animals) nor do all the quadrupeds lay eggs (for
horse and cow and countless others bring forth animals) nor do
20   they all bring forth animals (for lizards and crocodiles and many
others lay eggs). Nor does the difference lie in having or not having feet; for [some] footless [animals] bring forth animals, for example, the vipers and the selachians, but others lay eggs, for
example, the class of the fish and those of the other snakes. And
of those that have feet, many lay eggs and [many] bring forth animals, for example, the quadrupeds that have been mentioned.
25   And both bipeds, for example, human being, and footless [ani-
mals], for example, whale and dolphin, produce animals inside
themselves. So then it isn't possible to make the division in this
way, nor [is] any of the locomotive organs responsible for this
difference, but those of the animals that are more perfect in their
nature and that share in a purer ruling beginning bring forth ani-
30   mals; for nothing produces animals inside itself unless it draws
in breath and breathes. Now more perfect [are] those that are
hotter in their nature and wetter and not earthy. And the lung,
of all those whose [lung is] full of blood, is the determining mark
of the heat that is natural [to animals]; for in general those that
have a lung are hotter than those that don't have [one], and of
these themselves, those [are hotter] that don't have one that is
35   spongy or firm or poor in blood, but [one that is] full of blood
733a   and soft. And as the animal is perfect, whereas the egg and the
larva are imperfect, so the perfect naturally comes into being
from the perfect. And as for those that are hotter, on account of
having a lung, but drier in their nature, or those that are colder
5   but wetter, the ones lay a perfected egg, and the others, after producing an egg, produce animals inside themselves. For the birds

---

7   This is perhaps a reference to 715b4–16.

56

and the horny-plated [animals], on account of heat, perfect [what they bring forth], but on account of dryness they lay eggs, whereas the selachians, [being] less hot than these but more wet, [are of a nature] so that they share in both; for they produce eggs and they produce animals inside themselves, they produce eggs because [they are] cold, and they produce animals because [they are] wet; for the wet is conducive to life, whereas the dry is furthest from the ensouled. And since they have neither feathers nor horny plates nor scales, which [are] signs of a more dry and earthy nature, the egg [that] they generate [is] soft; for as in [the animal] itself, neither in the egg does what is earthy come to the surface. And on account of this they lay eggs inside themselves; for if the egg went outside, it would be destroyed for lack of protection. The [animals] that are both cold and more dry lay eggs, but the egg [is] unperfected, and [it is] also hard-skinned, on account of the [animals] being earthy and [on account of] its being released unperfected, so that it may be saved by having its earthenware-like [covering] as a protection. And so fish, being scaly, and the soft-shelled [animals], being earthy, generate eggs that have hard skins. And the cephalopods, just as they themselves [are] sticky in the nature of their body, so they preserve their eggs, which are released unperfected; for they release a lot of stickiness around the embryo. And all the insects bring forth larvae. All the insects are bloodless, wherefore also [we find them] bringing forth larvae externally.[8] But the bloodless [animals] don't all bring forth larvae in an unqualified sense; for the insects, namely, the [animals] that bring forth larvae,[9] and those

10

15

20

25

---

8 At 733a25–26, Drossaart Lulofs' text reads, instead of *dio kai skōlēkotokounta thuraze*, which I read, on the authority of a few manuscripts, including Z, the oldest, *dio kai ta skōlēkotokounta thuraze*, which appears in the preponderance of the manuscripts. His text would be translated, instead of "wherefore also [we find them] bringing forth larvae externally," as "wherefore also [are] those that bring forth larvae externally."

9 At 733a27, Drossaart Lulofs would delete the word *kai*, which I have translated here as "namely," though he has no ancient authority for doing so. He apparently thought that it would have to be translated in its more usual sense of "and," and didn't think that Aristotle could be making a distinction between the insects and the animals that give birth to larvae. But since the word *kai* can indeed mean "and in particular," or "namely," I see no reason for deleting it. I might also note that most of the manuscripts—though not Z—omit the entire phrase that I have translated as "namely, the [animals] that bring forth larvae."

that lay their eggs unperfected, for example, the fish that have scales and the soft-shelled [animals] and the cephalopods, over-
30   lap with one another. For the eggs of the latter are larva-like (for they increase in size externally), whereas the larvae of the former become egg-like as they develop; in what manner [this happens] we will determine in the later [speeches]. Now one must understand how well and consecutively nature produces coming into
733b   being. For the more perfect and hotter of the animals produce their offspring [already] perfected as regards what sort [they are] (as regards how big [they are], none at all of the animals [do so]; for [the offspring] all increase in size after they have come into being), and indeed these generate animals inside themselves im-
5   mediately. Those [that come] second don't generate perfected [offspring] inside themselves immediately (for they bring forth animals after having first produced eggs), but they bring forth animals externally. Others don't generate an animal [that is] perfected, but they generate an egg and this egg is [already] perfected. Others, whose nature is still colder than these, while they generate an egg, don't [generate] a perfected egg, but it is per-
10   fected externally, as for example the class of scaly fish and the soft-shelled [animals] and the cephalopods. And the fifth and coldest kind doesn't lay eggs out of itself, but even this sort of happening takes place externally to it, as has been said. For the insects bring forth larvae at first, but as the larva develops it be-
15   comes egg-like (for the so-called chrysalis has the capacity of an egg), then from this there comes to be an animal, reaching the end of its coming into being in its third change. Now some of the animals don't come into being from seed, as was said earlier;[10] but the blooded [ones] all come into being from a seed, as many as come into being from coupling, with the male emitting semen into
20   the female, from which, when it has entered, the animals are constituted and assume their own shape, some inside the animals themselves, as many as give birth to animals, but others in eggs and seeds and other such secretions.[11] Concerning which [things]

---

10 715a21–25; 715b26–30; 731b10–13; 732b11–12.

11 Drossaart Lulofs thinks that there must be a lacuna in the manuscripts after the word translated as "eggs." He notes that it is only in the case of plants that offspring assume their own shape within seeds and other such secretions.

there is a greater perplexity, [namely,] how ever does the plant or any one whatsoever of the animals come into being out of the seed. For [it is] a necessity that what comes into being come into being   25
out of something and by the agency of something and [that it become] something. Now [that] out of which is the material, the first of which some animals have inside themselves, having obtained it out of the female, as [is the case with] as many as are not brought forth as animals, but are brought forth [as] larvae or are brought forth [as] eggs, while others until [they are] far advanced take it out of the female through suckling, as do as many   30
as are brought forth as animals, not only externally but also internally. And so [that] out of which they come into being is the material of that sort. But [what] is sought now [is] not [that] out of which but [that] by the agency of which the parts come into being; for either one of the [beings] external to [the seed] produces them or [that which does so] is present[12] in the semen and seed, and this is either some part of soul or soul or it would be   734a
[something] that has soul. Now that any of the external [beings] makes each [one], either of the entrails or of the other parts, would seem unreasonable; for [it is] impossible to set [anything] in motion without touching [it], and for anything to be acted upon by this without its setting [it] in motion. Something is already present, then, in the embryo itself, either a part of it or   5
separate. Now [for it] to be something other [that is] separate [is] unreasonable; for after the animal has been generated is it destroyed or does it remain? But nothing of such a sort is manifestly present in [what has been generated], [nothing] that is not a part of the whole, either plant or animal. But surely also, for it to be destroyed after having produced either all the parts or some   10
[one of them is] absurd; for what will produce the remaining

The words translated as "and seeds and other such secretions" were apparently not included in the text used as the basis for the Arabic translation. But I am inclined to agree with Drossaart Lulofs that there is a lacuna in the manuscripts.

12   At 733b33, Drossaart Lulofs would read, instead of *ē enuparkhei*, the reading of the preponderance of the manuscripts, *ē enuparkhon ti*, which is based only on one of several editions of William of Moerbeke's Latin translation. Drossaart Lulofs' text would be translated, instead of as "or [that which does so] is," as "or something that is present."

ones? For if that [has produced] the heart, [and] then has been destroyed, and this [has produced] a different [part], it belongs to the same account that either they are all destroyed or that they all remain. It is preserved then; then there is a part of it [i.e., of what is coming into being] that is present straightaway in the

15 seed. And plainly, if there is nothing of the soul that is not in some part of the body, it would also be some part that is ensouled straightaway. The other [parts], then—how [are they produced]? For either all the parts come into being simultaneously, for example, heart, lung, liver, eye, and each of the others, or [they do so] successively, as in the so-called verses of Orpheus; for he as-

20 serts there that the animal comes into being in a way that is like the plaiting of the net. Now that they don't [come into being] simultaneously is manifest even to sense-perception; for some of the parts are manifestly present already [in the embryo], but others not. And that [it is] not on account of smallness [that] they don't appear [is] clear; for the lung, though it is bigger in size than the heart, appears later than the heart in the coming into

25 being from the beginning. And since one [part is] earlier and the other later, does the one produce the other and is it [present] on account of what is next to it or rather does this come into being after this? I mean, for example, that [it is] not the heart, when it has come into being, [that] produces the liver, and this something different, but that this [comes into being] after this, as after a boy a man comes into being, but not by the agency of that. And

30 the argument for this [alternative is] that in the [things] that come into being by nature or by art, that which potentially is comes into being by the agency of that which is in the complete sense, so that [on the other alternative] the form and the shape would have to be in that, for example, the [form] of the liver in the heart. And in other ways too the [other] account is absurd and fabricated. But besides, for there to be present in the seed

35 straightaway some part of the animal or plant, [a part] that has [already] come into being—whether it is able to produce the other [parts] or not—[is] impossible if everything comes into being out of seed and semen. For [it is] clear that it came into being by the agency of that which produced the seed, if indeed

734b it is present in [it] straightaway. But seed has to have come into being before, and this [is] the work of that which generates; it

isn't possible, then, for any part to be present [in it straightaway].
Therefore it doesn't have within itself that which produces the
parts. But surely [this] is not external to [it]; and [it is] a necessity
that it be one of these two. One must try, then, to resolve these
[things]; for perhaps something among the [things] that have   5
been said is not unqualified, for example, [let us ask] how ever is
it not possible [for the parts] to come into being by the agency
of what is external to [the seed]. For there is a sense in which it
is possible and a sense in which it is not. Now to say the seed or
[that] from which the seed [has come to be] doesn't make any
difference insofar as [the seed] has within itself the motion that
that one initiated. And it is possible for this to set this in motion,
and this this, and [for it] to be like the automata among the pup-   10
pets. For their parts are present, having somehow a potentiality,
when they are at rest; of which [parts], when one of the external
[beings] sets the first one in motion, the one next to it straightaway comes to be as an actuality. So then just as in the case of
the automata, in a certain manner that [being] sets [the parts] in
motion without touching anything now, but having touched; and
likewise also [that] from which the seed [has come to be] or that   15
which produced the seed [produces the parts], having touched
something, but not touching any longer; but in a certain [other]
manner the motion that is within [the seed produces them], just
as housebuilding [produces] the house. So then that there is
something that produces [the parts], but not [something that is]
like a particular this, nor that is present at first as [something]
perfected in [the seed], [is] clear. And from here [we] must grasp   20
how ever [it is that] each [part] comes into being, first taking [as]
a beginning that as many [things] as come into being by nature
or by art come into being through the agency of [something] that
is actually out of what is potentially of such a sort. Now the seed
is such, and it has a motion and a ruling beginning of such a sort
that, when the motion ceases, each of the parts comes into being
and [comes into being] ensouled. For it is not a face if it doesn't   25
have soul, nor flesh, but when they have perished the one will
be said to be a face, and the other flesh, in name only, just as if
they came into being of stone or of wood. And the uniform and
the instrumental [parts] come into being at the same time. And
just as we wouldn't say that the fire alone produced an axe or

30   another tool, so also not a foot or a hand. And in the same manner not flesh either; for there is a certain work belonging to this
as well. Now heat and cold could make [things] hard and soft
and tough and brittle and [with] as many other such attributes
as belong to the ensouled parts, but the defining character (*logos*)
by which this [is] already flesh and that bone, no longer, but
35   [what produces this is] the motion derived from the [being] that
did the generating, which [being] is in the complete sense what
[that] out of which they come into being is potentially, as also in
the case of the [things] that come into being in accord with art;
735a   for the hot and the cold make the iron hard and soft, but the motion of the tools, which has a rational character (*logos*), that of
the art,[13] [makes] a sword. For the art is a ruling beginning and
form of that which is coming into being, but in another; whereas
the motion of nature [is] in [the being] itself, being derived from
5   a different nature that has the form in actuality. And does the
seed have soul or not? [There is] the same account as concerning
the parts; for there will be no soul in [anything] else except in
that of which it is in fact [the soul], nor will there be a part that
doesn't share in [soul], except in name only, like an eye of a dead
[person]. [It is] clear, then, that it both has, and is, [soul] poten-
10   tially. And it is possible for it to be nearer to and further from
itself in [its] potentiality, as the sleeping geometer is further than
the waking one, and this one [further] than the one who is con-
templating. So then no part is responsible for this [process of]
coming into being, but [rather] the external first mover. For
nothing itself generates itself; but when it has come into being,
it is already causing itself to grow. Wherefore some [part] comes
15   into being first, and not all at the same time. And [it is] a necessity that this come into being first, [namely,] that which contains
the ruling beginning of growth; for whether plant or animal,
this, the nutritive, belongs alike to all. (And this is what is generative of another like itself; for this [is] the work of everything
[that is] by nature perfect, both animal and plant.) And [it is] a

---

13 At 735a2, Drossaart Lulofs would delete the definite article before the word
translated as "art," though without any ancient authority. According to his
text, the phrase translated as "a rational character (*logos*), that of the art"
would be translated instead as "a rational character (*logos*) of the art."

necessity on account of this that when something has come into   20
being, [it is] a necessity [for it] to grow. Therefore, while that
which is synonymous generated [it], for example, a human being
a human being, it grows on account of itself. Being itself something, then, it causes [its] growth. Now if [it is] some one [thing],
and this [is] first, [it is] a necessity that this come into being
first. So if the heart comes into being first in some animals, and
in those that don't have a heart, what is analogous to this, the
ruling beginning would be from this in those that have [it], and   25
in the others, from what is analogous. So then, on the one hand,
what is responsible as ruling beginning for the coming into being
concerning each [part], setting in motion first and fashioning,
has been stated in relation to the perplexities that were raised
earlier.

Chapter 2

Someone might, on the other hand, be perplexed concerning the
nature of seed. For the seed comes out of the animal thick and   30
white, but as it cools it becomes wet like water and with the color
of water. [This] might seem quite strange; for water isn't thickened
by [what is] hot, but that comes out thick from inside, out of [what
is] hot, and as it cools it becomes wet. And besides, [things] that
are watery freeze; but seed doesn't freeze when placed in frost out   35
in the open, but it becomes wet, suggesting that it had been thickened by the opposite. Yet neither [is it] plausible that it is thickened
by [what is] hot. For as many [liquids] as contain a higher degree   735b
of earth, these are [the ones that are] condensed and thickened by
being boiled, as for example milk. It ought, then, to be solidified
as it cools. But as it is, nothing [of it] becomes solid, but it all [be-
comes] like water. The perplexity, then, is this; for if it is water[14]—
water is manifestly not thickened by heat, whereas this comes out   5
thick and hot and out of the body that is hot; but if it is of earth
or mixed from earth and water, it shouldn't all become wet and

---

14 At 735b4, Drossaart Lulofs would emend, though without any authority, the
    manuscripts' *hudōr*, which I have translated as "water," with *hudatos*, which
    would be translated as "[composed] of water."

water.[15] Or have we not distinguished all the [things] that occur? For [it is] not only the wet [substance] constituted out of water and [what is] earthy [that] is thickened, but also that [constituted] out
10 of water and air (*pneuma*),[16] as for example foam becomes thicker and white, and the fewer and less distinct the bubbles are, the whiter and more firm does its bulk appear. And also olive oil is affected in the same way; for it is thickened when mixed with *pneuma*; where-
15 fore also that which becomes whiter becomes thicker, since what is watery in it is separated out by heat and becomes *pneuma*. And lead ore, when mixed with water and olive oil and pounded, makes a large bulk out of a little and [something] firm out of [what was] wet and white out of black. And responsible [for this is] that
20 *pneuma* gets mixed in, which produces the bulk and lets the whiteness show through, as in foam and snow; for snow too is foam. And water itself mixed with olive oil becomes thick and white; for *pneuma* gets enclosed by the friction, and olive oil itself contains much *pneuma;* for oiliness is [a characteristic] neither of earth nor
25 of water, but of *pneuma*. Wherefore also it floats on the surface of water; for the air that is in it as in a vessel carries [it] up and floats on the surface and is responsible for its lightness. And in periods of cold and frost olive oil thickens, but it doesn't freeze; for on
30 account of [its] heat it doesn't freeze (for air is hot and doesn't freeze), but on account of its [i.e., the air's] being compacted and condensed, olive oil becomes thicker by the agency of cold.[17] On account of these causes seed too comes out firm and white from inside [the animal], since it contains much *pneuma*, which is hot, through the agency of the heat inside, but after coming out, when
35 it has given off its heat and the air [in it] has been cooled, it be-

---

15 At 735b7, Drossaart Lulofs would delete the manuscripts' *kai hudōr* (which I have translated as "and water") or else replace them with words meaning "like water." Again, he has no authority for either of these changes.

16 Cf. Book One, n. 59.

17 At 735b31 the text I have translated as "by the agency of cold" appears to have been the basis for the Arabic translation. But this phrase is preceded in the manuscripts by the word *hōsper*, which means "just as," so that the whole phrase would be translated as "just as by the agency of cold." Drossaart Lulofs would delete the word *hōsper*, as I do in my translation, but some scholars have suggested instead that it was originally followed by a word or words that have been lost.

comes wet and dark; for water, and any earthy [stuff] that may
be mixed in, as in phlegm, are left also in the seed as it dries.
Semen, then, is compounded of *pneuma* and water, and *pneuma*   736a
is hot air; wherefore it is wet in its nature, because [it is consti-
tuted] out of water. For Ktesias of Knidos manifestly spoke
falsely in what he said concerning the seed of elephants. For he
asserts that when it dries it becomes so hard that it becomes like   5
amber. But this doesn't come about; for while [it is] necessary
that one seed be more earthy than another, and [that it be]
especially so in those [animals] in which much earthy [stuff] is
present in keeping with the bulk of their body, [it is] thick and
white on account of *pneuma* being mixed in. And indeed the seed
of all [animals] is white; for Herodotus doesn't speak truly in as-   10
serting that the semen of Ethiopians is black, as if were necessary
that everything belonging to those whose skin is black be black,
and this despite seeing that their teeth are white. And responsible
for the whiteness of seed [is] that semen is foam, and foam is
white, and especially that which is made up out of the smallest   15
parts and [out of parts] so small that each bubble is invisible,
which happens also in the case of water and olive oil when they
are mixed and [their parts] are rubbed together, as was said earlier.
And it looks as if it also didn't escape the notice of the ancients
that the nature of seed is foam-like; at any rate they named the   20
goddess[18] [who is] sovereign over intercourse after this power.
And so the cause of the perplexity that was mentioned has been
stated, and [it is] manifest that [it is] on account of this [that] it
also doesn't freeze; for air doesn't freeze.

Chapter 3

Next after this is to raise the perplexity and to say, if in the case of
the [animals] that emit semen into the female what enters is no part   25
of the embryo that is coming into being, where its bodily [substance]
goes, if indeed it does its work by means of the capacity that is present in it. And one must determine whether that which is being con-

---

18 This is a reference to Aphrodite, whose name is sometimes thought to be de-
    rived from the Greek word *aphros*, meaning "foam" or "sea foam."

stituted within the female receives something from what has entered
or nothing, and concerning soul in virtue of which it is called an
30   animal (it is an animal in virtue of the part of the soul that is sense
perceptive), whether it is present in the seed and the embryo or not,
and where [it is] from. For no one would count the embryo as soulless, deprived of life in every manner; for the seeds as well as the
35   embryos of animals are no less alive than plants, and [they are] fertile up to a point. So then [it is] manifest that they have the nutritive
soul (the [reason] on account of which it is necessary to acquire this
[soul] first [is] manifest from what (pl.) has been determined con-
736b   cerning soul in other [speeches]), and as they develop [they acquire]
also the sense perceptive [soul] in virtue of which [each is] an animal,[19] for it doesn't become an animal and a human being at the
same time nor an animal and a horse, and likewise in the case of the
other animals; for the end comes into being last, and what is peculiar
5   [to it] is the end of the coming into being of each. Wherefore also
concerning intellect, when and how those that have a share in this
ruling beginning acquire it, and from where, [this] involves a very
great perplexity, and one must try eagerly to grasp [it] according to
one's power and to the extent that it is possible. And so [it is] clear
that one must treat seeds and embryos that are separable[20] [from
10   the mother] as having the nutritive soul potentially, but not actually
until, like those of the fetuses that become separated, they draw in
nourishment and do the work of that sort of soul; for at first all such
[beings] seem to be living the life of a plant. And [it is] clear that

---

19   At 736b1, Drossaart Lulofs thinks that there is a lacuna in the manuscripts,
and he reports that in a later manuscript words were added to the margin that
would be translated "and the rational [soul] by virtue of which [it is] a human
being," which he suggests is more or less what Aristotle must have written.
At any rate, the "other [speeches]" referred to earlier in this sentence include
speeches from *On Soul*, Book Two, Chapters 2 and 3, and Book Three, Chapters 12 and 13.

20   At 736b9 Drossaart Lulofs reads, instead of *ta khōrista*, which appears in the
great preponderance of the manuscripts and which I translate as "that are
separable," *ta mēpō khōrista*, which appears only as a marginal addition in one
of the later manuscripts and which would be translated as "that are not yet
separated." Now it seems clear that Aristotle is speaking of seeds and embryos
that are not yet separated from the mother. But since the word *khōrista* can
mean "separable" as well as "separated," and since what is (merely) separable
is not yet separated, I choose to translate the manuscript reading.

one must speak in a corresponding way concerning the sense perceptive soul and the thinking soul; for [it is] necessary to have all   15
[these] potentially before [having them] actually. And [it is] necessary either [1] for them all, not being before, to come to be in [the
material], or [2] for them all, existing before, [to come to be in it],
or [3] for some [to come to be in it, existing before], but others not;
and for them to come to be in [it] either [1] not having entered the
material in the seed of the male or [2] having entered here from
there, with them coming to be in the male either [1] all from outside   20
or [2] none or [3] some but not others.²¹ Now then that [it is] not
possible for them all to exist before is manifest from the following
sorts of [considerations]; [regarding] all those ruling beginnings
whose actuality [is] bodily, [it is] clear that [it is] impossible for these
to exist without a body, as for example to walk without feet; so that
[it is] impossible for them to enter from outside; for neither [is it]   25
possible for them to enter themselves by themselves, since they are
inseparable [from body], nor for them to enter in a body; for the
seed is a residue of nutriment that is undergoing a change. It remains, then, for the intellect alone to enter from outside as an addition and for it alone to be divine; for bodily actuality has no share
in its actuality. Now then it looks as if the capacity of all soul is as-   30
sociated with a different and more divine body than the so-called
elements, though as souls differ from one another in preciousness
or lack of it, so also does this sort of nature differ. For in the seed of
all [ensouled beings] there is present that which makes the seeds

---

21 This sentence might also be translated as "And [it is] necessary either [1] for
them all, not being [in the material] before, to come to be in [it], or [2] for
them all [to be in it], being in [it] before, or [3] for some [to be in it, being in
it before], but others not; and [for those that *come to be* in it] to come to be in
[it] either not having entered the material in the seed of the male or [2] having
entered here from there, with them coming to be in the male either [1] all from
outside or [2] none or [3] some but not others." I'm not certain that my translation is better than this, but I have chosen it largely because another form of
the word that I translate at 736b17 as "existing before," but which is rendered
in the alternative version as "being in [it] before" (*proüparkhousas*), is used in
the following sentence as an apparent equivalent to "coming to be in [the male]
from outside" (*proüparkhein*, 736b21; cf. b19–20, b24), and thus not being "in
it" before, but rather existing somewhere else. And it doesn't seem likely to
me that Aristotle would have used the word in this sense if he had just used it
in the previous sentence to mean "being in [it] before."

35   fertile, that which is called hot. But this is neither fire nor a power of that sort but the *pneuma* that is enclosed in the seed and in [its] foaminess, i.e., the nature in the *pneuma*, which [nature] is analo-

737a   gous to the element of the stars. Wherefore fire doesn't generate any animal; and it is manifest that no [animal] is constituted in [things] subjected to fire, neither in wet nor in dry [ones]. But the heat of the sun and that of animals [do generate], not only the [heat conveyed] through the seed, but also if there happens to be some

5   different residue of their nature, likewise this too contains a life-giving ruling beginning. So then that the heat in animals is neither fire nor has its ruling beginning from fire is manifest from these sorts of [considerations]. But the body of the semen in which the seed also goes out, the [seed] consisting of the soul-ish ruling be-ginning ([it] being partly separable from body, in all those in which

10   something divine is encompassed—and of such a sort is what is called intellect—and partly inseparable), this [inseparable part of the] seed belonging to the semen[22] dissolves and evaporates (*pneu-matoutai*), since it has a wet and watery nature. Wherefore one should not always be looking for it to emerge outside, nor for it to be any part of the shape that has been constituted, just as also not for the fig juice that has caused milk to set; for this too changes and

15   is no part of the masses that are constituted. And so concerning soul, how embryos and semen have it and how they don't have it, has been determined; for they have it potentially, but they don't have it actually. And since seed is a residue and is moved with the same motion as that with which the body grows when the last [stage

20   of] nutriment is being divided up, when it enters the uterus, it con-stitutes the residue of the female and sets that too in motion with the same motion that it itself is in fact moved with. For that too is residue, and it has all the parts potentially though none actually. For

---

22   At 737a11, I read *touto to sperma tēs gonēs*, which is the reading of all the manu-scripts and which I translate as "this [inseparable part of the] seed belonging to the semen." Drossaart Lulofs, however, as well as all the other modern edi-tors I am aware of, reads *touto to sōma tēs gonēs* (substituting the word *sōma* for the word *sperma*), even though this is a modern emendation without any ancient support. On the basis of this emendation, instead of "this [inseparable part of the] seed belonging to the semen," the translation would be "this body of the semen." That reading is admittedly easier to interpret, but it seems to me that the manuscript reading is more likely to be what Aristotle wrote.

it even has potentially the sort of parts by which the female differs   25
from the male. For just as out of mutilated [animals] there sometimes come into being mutilated [offspring] and sometimes not, so
also out of a female there sometimes [comes into being] a female,
but sometimes not, but rather a male. For the female is as it were a
mutilated male and the menses are seed, but not pure [seed]; for one
[thing] alone they don't have, the ruling beginning of the soul. And   30
on account of this in as many of the animals as have wind eggs, the
egg that is being constituted has the parts of both [sexes] but it
doesn't have the ruling beginning, wherefore it doesn't become ensouled; for the seed of the male brings this in. And when it has a
share in such a ruling beginning, the residue of the female becomes
an embryo. There forms around wet, but body-like, [substances]   35
that are being heated—as also in the case of boiled [dishes] that are
being cooled—a crust. And [what] holds all bodies together [is] their   737b
stickiness; which [character], as [animals] develop and become bigger, is acquired by the nature of sinew, which [nature] holds the parts
of the animals together, being sinew in some and its analog in the
others. And of the same character (*morphēs*) are also skin and blood
vessel and membrane and the entire class of that sort; for these differ   5
by the more and less and in general by excess and deficiency.[23]

Chapter 4

As for those of the animals whose nature is more imperfect, when
there has come into being a perfect embryo but not yet a perfect
animal, it is expelled outside; on account of what causes, has been   10
said earlier.[24] It is perfect at the point when one of the embryos is
male and another female, in as many of the [animals] that come
into being as have this difference; for some generate neither a female nor a male, as many as don't themselves come into being out
of a female and a male nor out of animals that have intercourse.
We will speak later concerning the coming into being of these. But   15

---

23 Drossaart Lulofs thinks that this sentence, along with the two preceding, is an
    interpolation, which was perhaps transposed here from some other passage.

24 This may be a reference to the discussion in Book Two, Chapter 1, 732b26–
    733b16, though the causes in question are not clearly stated there.

as for those of the animals, the perfect ones, that are animal-bearing within themselves, until [the female] has generated an animal and sent it outside, it keeps the animal that is coming into being naturally united [with it] inside itself. And as for many as bring forth animals externally but first produce eggs inside themselves,

20 when the egg they have generated [is] perfect, the egg of some of these is detached, as [it is] of those that lay eggs externally, and the animal comes into being out of the egg inside the female, whereas [the egg] of some, when the nutriment in the egg has been used up, is perfected [with nutriment] from the uterus, and on account of this the egg is not detached from the uterus. The selachian-like fish have this difference [within the class], concerning

25 which [animals] one must speak later by themselves. But now one must begin first from the first [animals]; first are the perfect animals, and of that sort are those that are animal-bearing, and first of these [the] human being. Now then the secretion of seed comes about in all [animals] just like [that] of any other residue; for each

30 [residue] is moved into its own place without the breath (*pneuma*) forcing [it] at all and without [any] other such cause compelling [it], as some assert, asserting that the genitals draw [the seed] in like cupping-glasses, and with [the animals] forcing [it] by means of their breath, as if it were possible for either this residue or that from wet or dry food to have gone somewhere else if they hadn't

35 forced it; [they assert this] because [animals] assist their expulsions [from the body] with breath that has been collected together. But this is a common [feature] in connection with all [things] that have to be set in motion; for through holding one's breath the

738a [needed] strength arises, and yet even without this force residues are expelled, even in [animals] that are sleeping, if the places happen to be relaxed and full of residue. [What they assert] is like if someone were to assert that the seeds in plants are secreted by the

5 wind (*pneumatos*) on each occasion toward the places where they customarily bear fruit. But responsible for this, as has been said,[25] is that there are parts fit to receive all the residues, [parts] for the useless ones, such as the dry and the wet, as well as, for blood, what are called the blood vessels. Now in females, around the re-

10 gion of the uterus, with the two blood vessels, the great one and

---

25 Cf. 725a33–b4.

the aorta, being divided higher up, many thin blood vessels [pro-
ceeding from them] come to an end in the uterus; when these [ves-
sels] are overfull from nutriment, and with the [female's] nature
because of coldness not being able to concoct [it], it is expelled
through the thinnest blood vessels into the uterus, since they are
unable because of their narrowness to receive the excess of the   15
amount, and there comes about the event [that is] like a hemorrhage. Now in women the period is not precisely fixed, but it
wishes[26] to come about when the moon is waning, plausibly; for
the bodies of animals [are] colder when the environment happens
to become of that sort,[27] and the ends of the months [are] cold on   20
account of the departure of the moon, wherefore also it happens
that the ends of the months are stormier than the middles. And
so when the residue has changed into blood, the menses wish to
come about in keeping with the stated period, but when it hasn't
been concocted, some is continually secreted little by little; where-   25
fore [what are called] "the whites" come about in females when
they are still small, even children. Now both these secretions of
residues, if they are moderate, preserve their bodies, inasmuch as
there comes to be a purgation of the residues that are responsible
for their bodies being sick; but if they don't come about or if they   30
come about in excess, they do harm; for they produce either diseases or a diminishment of their bodies; wherefore also the whites,
if they come about continuously and are excessive, take away from
the growth of girls. And so from necessity females have this residue on account of the causes that have been stated; for since their
nature is not able to concoct, [it is] a necessity there come to be   35
residue, not only of the useless food but also[28] in the blood vessels,
and that it overflow when it spreads throughout the thinnest blood
vessels. But for the sake of the better and the end, nature appro-   738b

---

26 At 738a17, and again at a23, the Greek word *bouletai*, which I translate literally
    as "wishes," is more commonly understood in the weaker sense of "tends."

27 At 738a20, as a result of a misprint, Drossaart Lulofs' text reads *tosouton*,
    which means "of that size," rather than *toiouton*, "of that sort."

28 At 738a36, Drossaart Lulofs proposes adding here the words *tou haimatos*,
    meaning "of the blood," which were apparently present in the text used as
    the basis for the Arabic translation, even though they do not appear in any of
    the manuscripts.

priates it to this place for the sake of coming into being, so that it may become another [being] of the same sort as it was going [to be]; for it is already present, being potentially at any rate of the same sort as the body of which it is a secretion. So then [it is] nec-
5   essary that there come to be residue in all females, more in those with blood and, among these, most in [the] human being; but [it is] a necessity that in the others as well some accumulation collect in the region of the uterus. And what is responsible [for the fact] that [there is] more in those with blood and, among these, most in human beings has been stated earlier.[29] And for this sort of resi-
10 due being present in all females, but not in all males—for some don't emit semen, but just as those that emit [it] fashion the [being] that is being constituted out of the material in the females by means of the motion in the semen, so those of this sort do the same [thing] and constitute [another being] by means of the mo-
15 tion within themselves[30] in this part from where the seed is se- creted. This is the region around the diaphragm in all those that have [one]; for the heart or its analog [is the] ruling beginning of nature; what is below is an appendage and for the sake of this— responsible, then, for there not being generative residue in all
20 males but in all females [is] that the animal is an ensouled body. And the female always provides the material and the male [pro- vides] what does the fashioning; for we assert that each of the two of them has this capacity, and that for the one to be female and the other male [is] this. So that [it is] necessary for the female to provide body and bulk, but [it is] not necessary for the male [to
25 do so]; for [there is] no necessity that either the tools or the maker be present in what (pl.) is coming into being. The body, then, [comes] out of the female, whereas the soul [comes] out of the male; for the soul is the being of a certain body. And on account of this, in as many [animals] not of the same kind as have inter-

---

29 728a31–b21.

30 At 738a13, the manuscripts include the preposition *en*, which would change the meaning of this prepositional phrase, which I translate as "by means of the motion within themselves," to "in the motion within themselves." With support from the Arabic translation, which doesn't translate the word, Drossaart Lulofs suggests that it should be deleted, and my translation follows his lead, since I can't make sense of the manuscript reading.

course, female and male (and [those animals] have intercourse whose times [of heat] are equal and whose periods of gestation are close and whose bodies don't differ much in size), at first   30 there comes into being [an offspring] that, as far as resemblance is concerned, shares alike in both, for example those that come into being out of a fox and a dog, and a partridge and a chicken, but as time goes on, others coming into being out of others finally turn out in conformity with the female in shape, just as alien seeds [turn out] in conformity with the region [in which they take   35 root]; for this is what provides the material and the body for the seeds. And on account of this the part in females that receives [the seed] is not a channel, but the uterus has extension; whereas in those males that emit seed [there are] channels, and these [are]   739a bloodless. The residues, each [of them], are in their own places and become residue at the same time; before that nothing [be-comes residue], unless something [does so] with much violence and contrary to nature. And so the cause on account of which the generative residues in animals are secreted has been stated. When   5 the seed from the male of those that emit seed has entered, it so-lidifies the purest [part] of the residue—for in the menses as well, the greater [part] is useless, being[31] wet, just as the wettest [part] of the semen of the male [is]. And of the emission on a single   10 occasion,[32] the earlier [semen] is more infertile than the later in most cases; for it has less soul-ish heat because of its being un-concocted, whereas that which has been concocted has thickness and has become more a body. In all those, either among women or the other animals, in whom no external discharge comes about on account of there not being much useless residue in that sort   15

---

31 At 739a8, I accept Drossaart Lulofs' addition of the word *on*, which I translate here as "being," even though it is a modern emendation without any ancient authority. The letters *on* happen to be the same as the last two letters of the previous word in the Greek text, and so it is certainly conceivable that some early scribe omitted them through negligence.

32 At 739a9–10, Drossaart Lulofs punctuates the text so as to include the words that I translate as "And of the emission on a single occasion" at the end of the previous sentence, instead of at the beginning of this one. With that punctua-tion, the end of the previous sentence would be translated as "just as the wettest [part] of the semen of the male, that is, of the emission on a single occasion, [is]." And this sentence would begin with the words "The earlier . . . ."

of secretion, what comes to be within [them] is as much as what is retained in those animals that discharge externally, which is solidified by the capacity of the male in the secreted seed or, as manifestly happens in some of the insects, when the part analo-
20   gous to the uterus has gone into the male. That the wetness that comes into being along with pleasure in females doesn't contribute anything to the embryo has been said earlier.[33] It might be thought [that it does] chiefly because, just as in males, in women too there come to be at night what they call wet dreams. But this is no sign; for these come to be also in the young among
25   the males who, though they are about to, don't [yet] emit any [semen], or in those who still emit [semen but semen that is] infertile. So then without the emission from the male during intercourse [it is] impossible to conceive, and [also] without the secretion of the feminine [residues], [this] either having gone outside or being sufficient inside. However, [even] if the females
30   don't have the usual pleasure connected with this sort of intimacy, they do conceive if the region happens to be excited and the uterus to have come down near [it].[34] But for the most part [conception] happens in the former way, on account of the mouth [of the uterus] not being closed when there comes into being the secretion with which pleasure usually comes to be both for males
35   and for women; in this condition there is freer passage also for the seed of the male. And his emission does not come about inside [the uterus], as some suppose (for the mouth of the uterus
739b   [is] narrow), but in the front, where the female discharges the moisture that comes to be in some of them, there also does the male discharge, if he releases moisture.[35] Now sometimes it re-

---

33 727b33–728a1.

34 At 739a31, Drossaart Lulofs reads *katō*, with apparent support from the Arabic translation, instead of *eggus*, which appears in the preponderance of the manuscripts and which I translate as "near [it]." If one were to accept his reading, the translation would simply omit the words "near [it]," since *katō*, which means "down," is already present as a prefix in the verbal form that I have translated as "to have come down."

35 At 739b2, Drossaart Lulofs would delete the words *ean tis exikmasēi*, which I translate as "if he releases moisture," even though they appear in all the manuscripts. According to Drossaart Lulofs, the Arabic translation does not translate these words.

mains, having[36] this manner [of being],[37] but sometimes, if the uterus happens to be suitably disposed and hot on account of the purgation, it draws [it] inside. A sign [of this]: pessaries that are 5 wet when applied are taken out dry. And further, in as many animals as have their uterus by the diaphragm, such as a bird and the animal-bearing [ones] among the fish, [it is] impossible for the seed not to be drawn there but to enter [simply] by being emitted. The place pulls the semen in on account of the heat that is present. And the secretion and collection of the menses kindle heat in this part, 10 so that [it draws the semen in,] just as conical[38] vessels, when they have been washed out with [something] hot, draw water into themselves when their mouth is turned downwards. The drawing in comes about in this manner, and in no way does it come about as some say [it does], by means of the parts that are instrumental 15 in relation to intercourse. And it happens in a way [that is] also opposite to [what is said by] those who say that the woman also releases seed; for it happens [according to them] that the uterus, having released [it] outside, draws it back in again, if indeed it will mix with the semen of the male. But for it to come about in this way [is] superfluous, and nature does nothing superfluous. When 20 the female's secretion inside the uterus has been coagulated by the

---

36 At 739b3, Drossaart Lulofs reads *ekhousa* (feminine participle), an emendation of his own, instead of *ekhon* (neuter participle), a manuscript reading that I accept (even though it is not that of the preponderance of the manuscripts, which read the neuter plural *ekhonta*). Drossaart Lulofs presumably made his emendation on the grounds that the Greek word used in this context for moisture is feminine. But since the moisture released by the male is his seed, for which the Greek word is neuter, it seems reasonable to refer to it with a neuter participle. See also n. 37, below.

37 Also at 739b3, Drossaart Lulofs, along with other modern editors, reads *topon* instead of the manuscripts' *tropon*, even though it is an unsupported modern emendation. Drossaart Lulofs' text would be translated, instead of "having this manner [of being]"—which I understand to mean being moist or wet— "occupying this place." But Aristotle has already indicated that the wetness of the male seed is not a permanent character, but rather a temporary manner of being. Cf. 737a11–12; 739a7–10.

38 At 739b12 I hesitantly accept Drossaart Lulofs' reading of *kōnika*, even though it is merely a modern emendation. However, several of the manuscripts read *akonēta*, which means "sharpened" or "whetted," and Aristotle may have written this word to refer to the shape of the vessels in question.

semen of the male, whose action is similar to that of rennet upon milk—for rennet too is milk that has life-giving heat, [heat] that brings what is like together into one and coagulates [it], and semen
25 is related in the same way to the nature of the menses; for it is the same nature, [that] of milk and [that] of menses—so then as what is body-like is coming together, what is wet is separated out and membranes consisting of the earthy [portions] that are becoming dry come to encircle [it] all around, both from necessity and for the sake of something; for [it is] necessary that the outer extremities of [things] heated and of [things] cooled become dry, and
30 there is a need for the animal not to be in wet [surroundings] but [to be] separated. And some of these [extremities] are called membranes and some choria,[39] differing by the more and less; and these are present in the egg-layers and the animal-bearers alike. As soon as the embryo has been constituted, it acts in about the same way
35 as [seeds] that are sown. For also in seeds the ruling beginning, the first one, is within them; and when this has become distinct, after being in [them] potentially before, from this are sent forth the shoot and the root. And it is this latter by means of which it
740a takes its nourishment; for the plant needs growth. So also in the embryo, while all the parts in a certain manner are in [it] potentially, the ruling beginning is present in [it, being] furthest along the way. Wherefore the heart is first to be actually differentiated. And this [is] clear not only on the basis of sense perception (for it
5 happens in this way), but also on the basis of reason; for when that which has come into being has been differentiated from both [parents], it needs to manage itself itself, just like a son who has been sent away from his father's house. So that it needs to contain a ruling beginning from which also later the ordering of the body comes about for the animals. For if [the ruling beginning] will be
10 from somewhere outside and will later be in [it], not only would someone be perplexed as to the when, but because [it is] a necessity, when each of the parts is being separated, that there be present first this [ruling beginning] out of which both growth and motion belong to the other parts. Wherefore as many as say, like Democritus, that the outer [parts] of animals are differentiated
15 first and the inner ones later don't speak correctly, [but] as if [they

---

39 Cf. 745b35; *On the History of Animals* 561b31–562a6.

were speaking] of wooden or stone figures; for such [things] don't
have a ruling beginning at all, but the animals all have [one] and
have [it] inside. Wherefore at first the heart manifestly becomes
distinct in all the blooded [animals]. For this is the ruling beginning, both of the uniform and of the non-uniform [parts]. For this
already deserves to be spoken of as the ruling beginning of the ani-   20
mal, i.e., of the composite, at the point when it needs nutriment;
for surely [it is] that which is [that] grows. And the last [stage of]
nutriment for an animal [is] blood and what is analogous [to it],
and the blood vessels [are] the receptacle for these; wherefore the
heart is the ruling beginning also of these. This is clear from the
Histories and the *Dissections.* Now since it is potentially by this
point an animal, but imperfect, [it is] necessary [for it] to get its   25
nutriment from elsewhere; wherefore it uses the uterus and the
[female] that carries it just as a plant [uses] the earth, for getting
nutriment, until it has been perfected to the point of being already
an animal that potentially moves about. Wherefore out from the
heart nature first traced the two blood vessels;[40] and from these
have been separated off small blood vessels going to the uterus,
[which are] what is called the umbilical cord. For the umbilical   30
cord is a blood vessel, one in some of the animals, but more than
one in others. And around these [there is] a skin-like sheath, [also]
what is called the umbilical cord,[41] on account of the weakness of
the blood vessels needing preservation and shelter. The blood vessels are attached to the uterus like roots, through which the embryo
gets its nutriment. For [it is] for the sake of this [that] the animal   35
remains in the uterus, and not, as Democritus asserts, so that its
parts may be molded in conformity with the parts of the [female]
that is carrying [it]. For this [is] manifest in the case of the egg-   **740b**
layers; for those get their differentiation inside the eggs after hav-

---

40 At 740a28, Drossaart Lulofs reads *prōtas*, which appears as a revision in one
    manuscript, instead of *prōton*, the reading of the great preponderance of the
    manuscripts. According to his text, instead of "nature first traced the two
    blood vessels," the translation would be "nature traced the first two blood
    vessels."

41 At 740a32, Drossaart Lulofs, like most modern editors, would delete the
    words translated in this sentence as "[also] what is called the umbilical cord,"
    though they are present—without of course a word corresponding to my
    bracketed "also"—in all the manuscripts.

ing been separated from the womb. But someone might raise the perplexity, if blood [is] nutriment, but the heart comes into being first, being full of blood, and blood [is] nutriment,[42] and nutriment

5   [is] from outside, where did the first nutriment come in from? Or this [is] not true, that all [nutriment is] from outside, but straight-away, just as in the seeds of plants there is something of that sort, [namely,] that which appears milky at first, so also in the material of animals the residue from their formation is nutriment. So then growth comes about for the embryo through the umbilical cord,

10 in the same manner as for plants through their roots, also for the animals themselves when they have been separated from the nu-triment inside themselves; concerning which [things] it must be spoken later at the times that are appropriate for the speeches. Now the differentiation of the parts comes about not as some take it, on account of like being of a nature to be carried toward like

15 (for in addition to the many other difficulties that this account con-tains, it follows that each of the uniform parts comes into being apart [from the others], for example, bones by themselves and sinews and flesh by itself, if someone should accept this cause); but because the residue of the female is potentially of such a sort

20 as the animal [is] by nature, and the parts are in [it] potentially, though none actually, [it is] on account of this cause [that] each of them comes into being, and because that which is active and that which is passive, when they are in contact in the manner in which the one [is] active and the other passive (by "in the manner in which" I mean the how and where and when), straightaway the

25 one acts and the other is acted upon. Now then the female supplies material, and the male [supplies] the ruling beginning of the mo-tion. And just as the [things] that come into being by the agency of art come into being through its tools—but it is truer to say through their motion; for this is the actuality of art, while art is the shape of the [things] that come into being in [something]

30 other—so the power of the nutritive soul, just as in the animals themselves and the plants it later produces growth out of their nu-triment, using heat and coldness as tools (for the motion of that

---

42 At 740b4, Drossaart Lulofs, like most modern editors, would delete the words translated here as "and blood [is] nutriment," even though they are present in all the manuscripts.

[soul is] in these, and each [being] comes into being by a certain ratio [of them]), so also from the beginning it constitutes that which is coming into being by nature. For it is the same material by which it grows and out of which it is constituted at first, so that   35 also the power that acts [upon this], that from the beginning,[43] [is] the same [as that which produces growth]; but this [power] is greater. If, then, this is the nutritive soul, this is also the [soul or power] that generates, and this is the nature of each [being], being   741a present both in plants and in animals, in all [of them], whereas the other parts of the soul are present in some but are not present in others of the living beings.[44] Now in plants the female is not separated from the male; but in the animals in which it is separated, the female needs the male in addition.   5

Chapter 5

And yet someone might be perplexed as to what the cause is; if, as is the case, the female has the same soul and the material is the residue of the female, why does the female need the male in addition, and does it not, rather, generate [by] itself out of itself? Responsible [for this is] that the animal differs from the plant by sense perception; and [it is] impossible for there to be a face or a   10 hand or flesh or any other part unless a sense-perceptive soul is in it either actually or potentially and either in some [limited] sense or in an unqualified sense; for it will be like a corpse or a

---

43 At 740b36, I read *to ex arkhēs*, even though it appears in only one manuscript, rather than *tōi ex arkhēs,* with Drossaart Lulofs, who follows the reading of the great preponderance of the manuscripts. With Drossaart Lulofs' reading, this clause would probably be translated as "so that also the power that produces [growth is] the same as that from the beginning." But that doesn't make sense to me, since the "this" in the next clause would then have to refer to "that from the beginning," and yet from what follows it seems clearly to be the power that produces growth.

44 At 741a3, I follow Drossaart Lulofs, who along with all the other modern editors I know of, reads *zōntōn*, an emendation with ancient support only from the Arabic translation, instead of *zōōn*, which is the reading of all the manuscripts. On the basis of the manuscript reading, instead of "are not present in others of the living beings," the translation would be "are not present in others of the animals."

part of a corpse. If, then, the male is what is productive of this sort of soul, where the female and the male are separated [it is]

15   impossible for the female to generate an animal [by] itself out of itself; for what has [just] been said was[45] [what it is] to be a male; although that the perplexity [just] stated is reasonable, at any rate, [is] manifest from the case of the birds that bring forth wind eggs because the female has the capacity to generate, at least up to a point. Further, this too contains a perplexity, how anyone will as-

20   sert that their eggs are alive; for neither is it possible in the way that fertile eggs are (for there would come into being out of them an actually ensouled [being]) nor in the way that a piece of wood or a stone is. For there is destruction of a sort of these eggs too as of [things] that previously in a certain manner had a share in life. [It is] clear then that they have a sort of soul potentially. Of what sort, then, [is] this [soul]? [It is] a necessity surely [that it be] the

25   last. And this is the nutritive; for this is present equally in all animals as well as plants. Why then does it not bring to perfection the parts and the animal? Because they need to have a sense-perceptive soul; for the parts of the animals are not like [those] of a plant. Wherefore they need partnership with the male; for the male is separated in these [kinds]. Which indeed also happens;

30   for wind eggs become fertile if the male mounts [the female] at a certain opportune time. But concerning the cause of these [things] it will be determined later. But if there is some kind which is female and doesn't have a male separated [from it], it may be that this generates an animal out of itself. While this hasn't been ob-

35   served convincingly up until now at least, [some] within the class of fish make one hesitate; for no male of those that are called erythrinus has yet been seen, but females, and [ones] full of embryos, [have been]. However we don't yet have convincing evidence about these; but there are also in the class of fish [some that are]

---

45   The Greek word that I translate "was" is a simple imperfect form of the verb "to be." But it is sometimes used, as I think it is here, not for something in the past, but for a truth that has been established as the result of an earlier discussion. I am unaware, however, of any earlier passage where Aristotle has made this precise claim about the source of the sense-perceptive soul, let alone established its truth. On the other hand, he did claim at 737a27–34 and at 738b25–26 that the soul, without qualification, comes from the male. See also 716a5–6, as well as 716a14–15 and a20–21.

neither females nor males, for example, the eels and a certain kind   741b
of mullets around the marshland rivers. But in as many as have
the female and the male separated, [it is] impossible for the female
itself by itself to generate to the end; for the male would be in vain,
and nature makes nothing in vain. Wherefore in such [kinds] the   5
male always brings to perfection the coming into being. For this introduces the sense perceptive soul either through itself or through
its semen. And since the parts are present potentially in the material,
when there comes to be a ruling beginning of motion, just as in
the puppets that are automata, the "one thing after another" follows of itself; and what some of the [thinkers] concerned with nat-   10
ure wish to say, the "[like] is carried toward like," ought to mean
that the parts are set in motion not in the sense of changing place,
but in that of remaining and being altered in softness and hardness
and colors and the other differentiating features of the uniform
[parts], becoming actually what (pl.) they were present [as] being
potentially before. Now the ruling beginning comes into being   15
first, and this is the heart in the blooded [animals] and what is analogous in the others, as has often been said. And this [is] manifest
not only through the sense perception, that it comes into being
first, but also in connection with the [animal's] ending; for to be
alive departs last from there, and it happens in all [cases] that what
came into being last departs first, and the first last, just as if nature   20
is running the return leg and retracing its steps toward the beginning from which it came. For coming into being is from what is
not to what is, while destruction is back again from what is to what
is not.

Chapter 6

After the ruling beginning the inner [parts] come into being before   25
the outer [ones], as has been said. And those that have [greater] size
appear before the smaller [ones], although some [of them] don't
come into being before. First the [parts] above the diaphragm are
articulated and are greater in size; what is below [is] both smaller
and less defined. And this comes about in all the [animals] in which   30
what is above and what is below are distinguished, except in insects;
but in those of these that are brought forth as larvae, growth comes

about toward what is above. For what is above is smaller at the beginning. Among the [animals] capable of locomotion, only in the
cephalopods are what is above and what is below without distinc-
35   tion. And what has been said is found also in the case of plants, that
the upper cavity precedes the lower in the [process of] coming into
being; for the seeds send out the roots before [sending out] the
shoots.⁴⁶ The parts of the animals are made distinct by means of
pneuma,⁴⁷ yet not by the [*pneuma*, i.e., breath] of the [female] that
generated [the animal], nor by its own, as some of the [thinkers]
742a   concerned with nature assert. This [is] manifest in the case of birds
and fish and insects; for some, after having been separated from the
[female] that generated [them], come into being out of an egg in
which they get their articulation, while some of the animals don't
breathe at all, but are brought forth as larvae or laid as eggs; and as
5   for those that breathe and get their articulation the womb, they
don't breathe until the lung has gotten its perfection; and both this
and the parts that precede it are articulated before they breathe.
Furthermore, as many of the quadrupeds as have multiple toes, for
10   example, dog, lion, wolf, fox, jackal, all generate blind [offspring],
and the eyelid is divided later after they have come into being. So
that [it is] clear that, in the same manner with regard to all the other
[parts] as well, just as their [being] of such a sort, [so] also their
[being] of such an amount comes to be, being present potentially
before and actually later, by the same causes as those by which their
[being] of such a sort is made distinct, and there come to be two
15   out of one. And [it is] necessary that *pneuma* be present, because [it
is] wet and hot, with this [i.e., the hot] acting and that [i.e., the wet]
being acted upon. Now some of the ancients who spoke about nature tried to say which of the parts comes into being after which,
though without being much acquainted with what (pl.) happens.
For among the parts, just as with regard to the other [things], the
20   one is naturally before the other. But "before" is already [spoken]
in many ways; for that for the sake of which and what is for the sake
of this differ, and while the one of them is before in the [process of]

---

46   In Aristotle's view the upper part of a living being is understood as that where
the nourishment is taken in, so that in plants it is the roots rather than the
stem (cf. *On Soul* 416a2–5; *On the Parts of Animals* 686b32–687a2).

47   Cf. Book One, n. 59.

coming into being, the other [is before] with regard to being. And
what is for the sake of this is also of two different sorts; for one is
where the motion [is] from, and the other [is] what that for the sake
of which uses. I mean, for example, that which is generative [of]
and that which serves as an instrument for what is being generated;   25
for the one of these needs to be present before, [namely,] the productive, for example, that which teaches [before] that which is learning, but flutes [need to be present] after the one learning to play the
flute; for [it would be] superfluous for flutes to belong to those who
don't know how to play the flute. And with there being three
[things]—one, the end, which we say is [that] for the sake of which,
and second, among the [things] for the sake of this, the ruling beginning that sets in motion and [is] generative (for the productive   30
and the generative, insofar as [they are] such, are relative to what is
being produced and what is being generated), and third, that which
is useful and which the end uses—[it is] necessary that there be
present, first, some part in which [is found] the ruling beginning
of the motion (and indeed this part is straightaway one [part] of the
end, and [the] most sovereign [one]), then after this, the whole and   35
the end, and third and last, the parts that serve as instruments for
these with a view to certain uses. So that if something of such a sort
is the very [thing that is] necessary to be present in animals,   742b
[namely,] that which contains the ruling beginning and the end of
all of [their] nature, [it is] necessary that this come into being first,
insofar as it sets in motion, [that it come into being] first, but insofar
as it is part of the end, along with the whole. So that of the parts
that serve as instruments, as many as are generative in their nature
must themselves always be present before (for they are for the sake   5
of another as a ruling beginning [from which it is generated]), but
as many of those for the sake of another as are not of that sort [must
come] later. Wherefore [it is] not easy to distinguish which of the
parts [come] before, as many as [are] for the sake of another or [that]
for the sake of which these [are]. For the parts that set in motion
intrude themselves, being before the end in [the process of] coming
into being, and [it is] not easy to distinguish those that set in motion   10
from those that serve as instruments. And yet [it is] along this path
of inquiry that one needs to seek what comes into being after what;
for the end, while later than some, [is] before others. And on account
of this, there comes into being first the part that contains the ruling

beginning, then, coming next, the upper trunk. Wherefore the
15 [parts] around the head and the eyes appear biggest at first in embryos (*embruois*), whereas those below the navel, for example, the
legs, [are] small. For [the parts] below [are] for the sake of that which
is above, and [are] neither parts of the end nor generative of it. But
as many as say that [animals] always come into being in such a way,
and hold this to be [the] ruling beginning in them, do not speak
finely nor [do they state] the necessity of the "on account of what,"
20 as for instance Democritus of Abdera, [who says] that there is no
ruling beginning of what is always and infinite, but that the on account of what [is] a ruling beginning and the always [is] infinite, so
that to ask for the on account of what concerning any of such [things]
is to seek, he asserts, a ruling beginning of the infinite. And yet according to this argument, according to which they demand that one
25 not seek the on account of what, there will be no demonstration of
any of the eternal [things]; but there manifestly is [demonstration]
of many, some always coming to be and some [always] being, since
that a triangle has [angles] equal to two right [angles is] always, and
that the diameter is incommensurable with the side [is] eternal, but
nevertheless there is something responsible for them and a demon-
30 stration. So then [that one ought] not to demand a search for a ruling
beginning of all [things] is said finely, but [that one ought not to demand this] of the [things] that are always and [always] come into
being, all [of them, is] not [said] finely, but [only with regard to] as
many of the eternal [things] as are in fact ruling beginnings [them-
selves]; for of the ruling beginning [there is] another [way of] gaining knowledge, and not demonstration. [The] ruling beginning in
the unmoved [things] is the what, but in those that come to be, there
are already more than one—but in a different manner [from one
another] and not all in the same one—of which one in number [is]
35 where the motion is from. Wherefore all the blooded [animals] have
a heart first, as was said at the beginning;[48] and in the others, what
743a is analogous to the heart comes into being first. Extending from the
heart [are] the blood vessels, like [the work of] those who draw
skeleton models on walls; for the parts are around these, inasmuch
they come into being out of these. The coming into being of the
5 uniform [parts] is by the agency of cold and heat; for they are con-

---

48 Cf. 740a3–4.

stituted and solidified, some by cold and some by hot. Concerning
the difference between these it has been spoken earlier in other
[speeches],[49] what sort of [bodies] are dissolvable by wet and fire
and what sort are undissolvable by wet and unmeltable by fire. Now
then the nutriment that oozes through the blood vessels and the
channels in each [of the parts], just like water in unbaked earthen-   10
ware, becomes flesh or what is analogous to this, being constituted
by the cold, wherefore also it is dissolved by fire. But as many of the
[incipient parts] springing up [as are] too earthy, having little wetness and heat, these, when they are cooled, as the wet evaporates
along with the hot, become hard and earthy in character, for example, nails and horns and hoofs and beaks; wherefore, though they   15
are softened by fire, none is melted, though some [are dissolved] by
what (pl.) is wet, for example, the shells of eggs. Sinews and bones
come into being by the agency of the internal heat, as the wetness
dries up. Wherefore bones are also undissolvable by fire, like earthenware; for as if in an oven, they have been baked by the heat during   20
their coming into being. But this [heat] does not make just anything
into flesh or bone nor at just any place nor at just any time, but
[only] that which is of a nature [to become these] and where it is of
a nature [to do so] and when it is of a nature [to do so]. For neither
will that which is potentially be [actually] through the agency of
that which can set in motion if [this] doesn't have its actuality, nor
will that which has the actuality produce out of just anything, as the   25
carpenter, for instance, couldn't produce a chest if not out of wood,
nor without him will there be a chest out of the wood. In the spermatic residue heat is present, having motion and actuality of such
an amount and of such a sort as is proportionate to each of the parts.
And to the extent that it falls short or is excessive, it will render that   30
which comes into being either inferior or deformed, in about the
same way as the [things] outside [the womb] that are coagulated
through boiling for the enjoyment of food or for some other busi-
ness. But here we provide the proportionateness of the heat with a
view to the motion,[50] whereas there the nature of the [male] that is

---

49 Cf. *Meteorology* 382a22–384b24 (Book Four, Chapters 5–7).

50 The Greek word translated as motion (*kinēsis*) is used by Aristotle not only
    for local motion, or change of place, but also for alteration, or change of qual-
    ity, increase and decrease, or change of size, and even coming into being and

35   generating supplies [it]. And for those [animals] that come into
     being spontaneously, the cause, from the season, is [its] motion and
     heat. Coldness is a lack of heat. And nature uses them both, [them]
     having, on the one hand, from necessity a capacity such that the
743b one produces this and the other that; however, in the [beings] that
     come into being, it happens for the sake of something that the one
     of them cools and the other heats and each of the parts comes into
     being, the flesh soft, [with them] making it such in one way from
5   necessity but in another way for the sake of something, and sinew
     dry and stretchable, and bone dry and breakable into fragments.
     The skin comes into being as the flesh dries, just as what is called
     the scum on boiled [liquids does]. And its coming into being hap-
     pens not only on account of its being outermost, but also because
     what is viscous is on the surface on account of its not being able to
10   evaporate off. Now in the other [animals] what is viscous is dried
     out (wherefore the outermost [parts] of the bloodless animals are
     hard-shelled or soft-shelled), but in the blooded ones what is vis-
     cous is more oily. And in all those among them whose nature is not
     too earthy, what is like lard collects under the covering of the skin,
15   which suggests that the skin comes into being out of viscosity of
     such a sort; for what is oily has a certain viscosity. And all these, as
     we said, must be said to come into being in one way from necessity
     but in another way not from necessity but for the sake of something.
     Now then the upper trunk is first separated off in the course of com-
     ing into being, but as time goes on the lower [part] takes its growth
20   in the blooded [animals]. And all [the parts] are distinguished in
     their outlines before, and later on they get their colors and their
     softnesses and their hardnesses, absolutely as if they were being
     fashioned by a painter, [namely,] nature; for painters too, by having
25   made a sketch with lines, in this way fill in the figure with its colors.
     Now because of the ruling beginning of the senses being in the
     heart, this also [is the] first [part] of the whole animal [that] comes
     into being, but because of the heat that belongs to this, where the
     blood vessels come to an end above, [there] the cold constitutes the

---

perishing, though sometimes, when speaking most strictly, he distinguishes
coming into being and perishing from motion proper (cf. *Physics* 201a9–15,
225a20–b9). In this context, then, I might have translated the word *kinēsis* as
"change," rather than "motion."

brain as a counterweight to the heat around the heart. Wherefore
the [parts] around the head get their coming into being next after   30
the heart, and they surpass the others in size; for the brain is large
and wet from the beginning. But what happens concerning the eyes
of animals involves a perplexity; for while they appear very big from
the beginning, in walking and swimming and flying [animals], they
come into being[51] last among the parts; for in the meantime they   35
collapse. Responsible [for this is] that the sense organ of the eyes,
like the other sense organs as well, [is set] upon channels; but while
the [organ] of touch and of taste is immediately either [the] body   744a
or something belonging to the body of the animals, and smell and
hearing [are] channels connecting to the air outside, full of inborn
pneuma, and terminating at the small blood vessels around the brain
that extend from the heart, the eye alone of the sense organs has a   5
body peculiar to it. It is wet and cold and isn't present beforehand
in its place—as [are] the other parts potentially, then coming into
being actually later—but from the wetness around the brain the
purest [part] is secreted through the channels that manifestly lead   10
away from them toward the membrane surrounding the brain. Evidence of this: while no other part in the head is wet and cold besides
the brain, the eye is cold and wet. From necessity then the region
gets big at first but collapses later. For it also happens in the same   15
manner in the case of the brain; for [it is] wet and large at first, but
as evaporation and concoction proceed, both the brain[52] and the
magnitude that is the eyes become more bodily and collapse. From
the beginning the head is very big on account of the brain, and on
account of the wetness in the eyes, the eyes show themselves big.   20

---

51 At 743b34, Drossaart Lulofs writes the word *sunistantai*, an emendation of
    his own, rather than *gignontai*, the reading of all the manuscripts. According
    to his text, instead of "they come into being," the translation would read "they
    are constituted." Now in either case, some adverbial notion such as "to per-
    fection" has to be understood. But this thought seems no more difficult with
    the manuscript reading than with Drossaart Lulofs' emendation.

52 At 744a18, most of the manuscripts contain the words *kai ta sōmata*, meaning
    "and the bodies" after the words translated here as "both the brain." But
    Drossaart Lulofs would omit them, following the lead of one of the older
    manuscripts as well as the Arabic translation, and I have chosen not to trans-
    late them. If they were included, I would have translated the end of this sen-
    tence as "the brain and the bodies and the magnitude of the eyes become more
    bodily and collapse."

But they are the last to attain their end because of the brain too being constituted [only] with difficulty; for it is late to give up its coldness and its wetness, in all the [animals] that have [one], and most of all in human beings. Indeed, [it is] on account of this [that] the bregma[53] is the last of the bones to come into being; for when

25   the fetuses (*embruōn*) have already come into being externally, this bone is soft in small children. And responsible for this happening most of all in human beings [is] that their brain is the wettest and largest among the animals, and responsible for this [is] that the

30   heat in their heart is the purest. The intellect makes clear their good temperament; for [the] human being is the most intelligent of the animals. But small children lack control over their head for a long time on account of the weight around the brain. Similarly also [they lack control] over as many parts as they need to set in motion; for the ruling beginning of motion is late to control

35   the upper [parts], and last [of all],[54] as many as whose motion, like [that of] the legs, is not closely connected with it. And the eyelid is such a part. And since nature makes nothing superfluous or in vain, [it is] clear that neither [does it make anything] too late or too early; for [if so,] what has come into being will be either

744b   in vain or superfluous; so that [it is] a necessity that the eyelids be separated at the same time as [the animal] is capable of setting [them] in motion. So then the eyes in animals are perfected late on account of the large amount of concoction [required] around the brain, and last on account of its being [only] when the motion is very much in control that it sets in motion those of the parts

5   that are both so far from the ruling beginning and [so] cooled. And the eyelids make it clear that they have such a nature; for if there is any amount of heaviness around the head on account of sleep or drunkenness or some other of such [conditions], we are unable to lift our eyelids, even though the weight they have is so

---

53   Aristotle understands the bregma as the front portion of the skull (cf. *On the History of Animals* 491a31–33).

54   At 744a34, I read *teleutaion*, with the great preponderance of the manuscripts, instead of Drossaart Lulofs' *teleutaiōn*, which is supported by only one manuscript and, as it appears to him, by the Arabic translation. Drossaart Lulofs' text would be translated, instead of as my "the upper [parts], and last [of all]," as "the upper and last [parts]."

small. And so it has been spoken concerning the eyes, how they come into being and on account of what, and on account of what cause they get their articulation last. Each of the other parts comes into being out of nutriment, those [that are] most precious and that have a share of the most sovereign ruling beginning, out of the nutriment that has been concocted and [is] purest and first, while the parts [that are] necessary and for the sake of these, out of the inferior [nutriment] and the leavings and residues. For like a good household manager, nature too is accustomed to throw away nothing out of which (pl.) it is possible to make something useful. In the management of households the best of the nutriment that comes into being is assigned to the freemen, while that which is inferior, and the residue from this, [is assigned] to household slaves, and what (pl.) is worst they give also to the animals that feed along with [them]. Just as the external intellect, then, turns these into growth, so in the [beings] that are coming into being themselves nature constitutes flesh and the bodies of the other sense organs out of the purest material, while out of the residues [it constitutes] bones and sinews and hair and, further, nails and hooves and all such [parts]; wherefore these are the last to get their constitution, when residues of the [animal's] nature are already coming into being. So then the nature of the bones comes into being in the first constitution of the parts out of the spermatic residue, and as the animals are growing they obtain their growth out of the natural nutriment out of which the sovereign parts [also do], but [out of] the leavings and what (pl.) is residue-like of this very [nutriment]. For there comes into being in everything the first [kind] and the second [kind] of nutriment, the one nutritive and the other growth-producing. Nutritive [is] what provides being to the whole as well as to the parts, while growth-producing [is] what brings about their increase in size. Concerning which [things] it must be determined more [fully] later. In the same manner as the bones, the sinews are constituted as well, and out of the same [things], out of the spermatic and the nutritive residue. But nails and hair and hooves and horns and beaks and the spurs of birds and any other part that there may be of such a sort [are constituted] out of the supplementary and growth-producing nutriment, which is obtained in addition from the female, i.e., from the [mother] outside [the growing

5   fetus].[55] And on account of this the bones obtain growth [only] up to a certain point; for there is a certain limit of size for all animals, wherefore also [there is a certain limit] of the growth of the bones. For if these had growth continually, as many of the animals as have bone or what is analogous would also grow as long as they lived; for these are the determining limit of size for animals. Now on account 10   of what cause these don't obtain growth continually must be said later; but hair and the [parts] akin to this grow as long as they are present, and more in diseases and when the bodies are growing old and wasting away, on account of more residue being left over, since less is being spent on the sovereign [parts] on account of old age 15   and diseases, although when even this [residue] fails on account of age, the hair also fails. But with the bones the opposite [happens]; for they waste away along with the body and its parts. But the hair even of those that have died [still] grows, although it doesn't, at any rate, come into being from the beginning. Concerning teeth someone might be perplexed. For they are possessed of the same nature 20   as the bones and they come into being out of the bones, whereas nails and hair and horns and the like [come into being] out of the skin, wherefore they also change colors along with the skin; for they come to be white and black and all sorts [of colors] in keeping with the color of the skin, but the teeth not at all; for they are [produced] out of the bones, in as many of the animals as have teeth and bones. 25   And they alone among the bones grow throughout life; this [is] clear in the case of the teeth that incline away from contact with one another. Responsible for their growth, in the sense of [its being] for the sake of something, [is that it is] on account of the work [they do]; for they would quickly be worn down if there didn't come to be some means of saving them, since even now in some [animals], 30   those that eat a lot but don't have big [teeth], when they grow old [their teeth] are wholly worn down; for they are being diminished at a greater rate than their growth. Wherefore nature has well con-

---

55 At 745a4, Drossaart Lulofs would delete the definite article *tēs* before the word *thurathen*, though without any ancient support. According to his emended text, the translation would read, instead of "from the female, i.e., from the [mother] outside [the growing fetus]," "from the female and from outside." But since Aristotle is speaking in this sentence of the original constitution of these parts, and not their subsequent growth, the manuscript reading, which refers only to the period of gestation, makes more sense to me.

trived this too with a view to what happens; for it brings the loss of the teeth together with their old age and their ending. But if their life were ten thousand or [even] a thousand years long, their [teeth] would have had to come into being extremely big in the beginning   35 and [would have had] to spring up many times; for [otherwise,] even if their growth had been continuous, they would nevertheless have   745b been ground down and been useless for their work. So then that for the sake of which they obtain their growth has been stated; but it is also a fact that the teeth don't have the same nature as the other bones; for those all come into being in the first constitution, and none later, but the teeth [come into being] later. Wherefore they are   5 able to spring up again after having fallen out; for they are in contact with, but aren't naturally united with, the bones. However, they come into being out of the nutriment that is distributed to the bones (wherefore they have [in a sense] the same nature), even then when those already have their own [full] number. Now then the other animals come into being, unless something contrary to nature comes   10 about, having teeth or what is analogous to the teeth, on account of their being released from the [process of] coming into being more [fully] perfected than the human being. But the human being, unless something contrary to nature happens, [is born] not having [them]. On account of what cause some of the teeth come into being and fall out, while others don't fall out, it will be said later. Because these   15 sorts of parts are [constituted] out of residue, because of this a human being has the least hair on its body of all the animals and it has the smallest nails in relation to its size; for it has the smallest amount of earthy residue. [It is] what is unconcocted [that] is residue, and what is earthy in the bodies of all [the animals is] the most   20 unconcocted. So then it has been said how each of the parts is constituted and what [is] responsible for [their] coming into being.

Chapter 7

Those of the fetuses (*embruōn*) that are born as animals have their growth, as was said earlier,[56] through the attachment of the umbilical cord. For since in the animals there is also the nutritive capacity of   25

---

56 740a24–35.

the soul, they immediately send out the umbilical cord, like a root, to the uterus. The umbilical cord is blood vessels in a sheath, more [of them] in the bigger [animals], for example, a cow and those of that sort, two in the medium-sized, and one in those at the [other] extreme.[57] Through this they get their nutriment in the form of blood;
30   for the uterus is the end point of many blood vessels. All the [animals] that don't have front teeth in both jaws, and of those that do have front teeth in both jaws, as many as whose uterus doesn't have one big blood vessel extending throughout [it], but instead of one, many [that are] close-set, these have in their uterus the so-called cotyledons,[58] to which the umbilical cord is connected and attached; for the blood vessels through the umbilical cord extend this way and that and branch out everywhere over the uterus; and where they terminate, there the cotyledons come into being,[59] having their rounded [side] toward the uterus and their hollow [side] toward the fetus (*embruon*).
35   Between the uterus and the fetus (*embruou*) are the chorion and the
746a   membranes. As the fetus (*embruou*) grows and is being perfected, the cotyledons become smaller and they finally disappear when it has become perfected. For nature sets out beforehand the nutriment in the form of blood for the fetuses (*embruois*) in this [part] of the uterus, as
5   in breasts, and on account of its accumulating little by little out of many [secretions], the body of the cotyledon becomes like a pustule

---

57 At 745b28, Drossaart Lulofs emends the manuscript reading, *tois eskhatois*, which I have translated as "those at the [other] extreme," so as to read *tois elakhistois*, which would be translated "the smallest." But I see no reason for an emendation of the text in a case such as this one where the meaning is clear without it.

58 The Greek word *kotylēdōn*, translated as "cotyledon," referred originally to the cup-like suckers on the arms of an octopus (*kotulē* means "cup"), and the term was later applied to pits in the inner wall of the uterus, which Aristotle is speaking of here.

59 Drossaart Lulofs places the words translated as "to which the umbilical cord is connected and attached; for the blood vessels through the umbilical cord extend this way and that and branch out everywhere over the uterus; and where they terminate, there the cotyledons come into being" within angular brackets, presumably because Immanuel Bekker didn't read them in the text of his 1831 edition, from which the standard page and line numbering is still derived. Though they are absent from most of the manuscripts, including Z, they do appear in several manuscripts, and are further supported by the Arabic translation.

or an inflammation. So then as long as the fetus (*embruon*) is smaller,
unable to take much nutriment, [the cotyledons] are clear and bigger,
but they collapse when it has grown. Most of the animals that are   10
mutilated [i.e., that lack horns,] and have front teeth in both jaws
don't have cotyledons in their uterus, but the umbilical cord extends
to [meet] a single blood vessel, and this is big and extends throughout
the uterus. And although some of these sorts of animals bear one [at
a time], whereas others bear many, those of the fetuses (*embruōn*) that
are more than one have the same manner [of receiving nutriment] as
the one. (One must contemplate these [things] from the illustrations
in the *Dissections* and from those drawn in the *Histories*.) For the ani-   15
mals are natural outgrowths [each] from its umbilical cord, and the
umbilical cord from the blood vessel, [with them being] next to one
another, as if along a water-pipe, [along] the flowing blood vessel. And
around each of the fetuses (*embruōn*) there are its membranes and its
chorion. Those who say that small children are nourished in the
uterus through sucking on some bit of flesh don't speak correctly;   20
for the same thing would happen in the case of the other animals, but
as it is, it manifestly doesn't (for [it is] easy to contemplate this
through the *Dissections*), and surrounding all the fetuses (*embrua*)
alike, those that fly and those that swim and those that go on land,
there are thin membranes that separate [them] from the uterus and   25
from the liquids that come to be in it, in which [membranes] themselves there is nothing of such a sort [as to be sucked on] nor is it
possible through these to procure the benefit of anything. And as for
those that are laid as eggs, [it is] manifest that they all obtain growth
outside, after having been separated from the womb, so that those
who speak in the way Democritus does don't speak correctly.[60] Coupling in animals comes about according to nature between those of

---

60 Drossaart Lulofs places the words translated as "so that those who speak in
     the way Democritus does don't speak correctly" within angular brackets, again
     presumably because Bekker didn't read them in the text of his 1831 edition.
     These words appear in only one of the manuscripts, to which Bekker appar-
     ently had no access, but they are also supported by the Arabic translation, and
     I accept Drossaart Lulofs' decision to include them, on the grounds that Aris-
     totle might well have wanted to identify at least one of the thinkers whose
     views he is criticizing here. For evidence that Democritus did in fact hold this
     view, see the reference in the textual apparatus of Drossaart Lulofs' edition
     to reports by two (admittedly much later) ancient physicians.

30 the same kind; nevertheless, [it comes about] also between those that have a closely related nature but that are not without a difference in species (*eidei*), if their sizes are about the same and their times of gestation are equal. Now such [things] are rare in the case of the other [animals], but they come about in the case of dogs
35 and foxes and wolves; and the Indian dogs are generated from
746b some wild dog-like [animal] and a dog. And this has also been seen happening in the case of birds that copulate frequently, for example, in the case of partridges and hens; and of the crooked-taloned [birds], hawks that differ in species are thought to have intercourse with one another. And it is in the same manner in the case
5 of some other birds. In the case of the sea animals nothing worth speaking of has been seen, but especially the so-called *rhinobati* are thought to come into being from a *rhinē* and a *batos* coupling.[61] And also, what is proverbially said about Libya, how Libya is always breeding something new, this is said to have been said on account of the fact that even those that are not of the same stock as one another have intercourse; for [it is said that] on account of
10 the scarcity of water, with [the animals] all heading for the few places that have springs, even those not of the same kind have intercourse. Now then the others of those [animals] that come into being from such intercourse manifestly couple again with one another and have intercourse and are able to generate the female as
15 well as the male, but mules, alone of such [animals], are sterile; for they do not generate, either from one another or by having intercourse with other [kinds]. Now there is the problem in general on account of what cause either a male or a female is sterile; for there are both women and men [who are] sterile, and [also some] in each of the kinds of the other animals, for example, horses and sheep.
20 But this kind, that of mules, is sterile as a whole. In the case of the others, the causes of sterility are found to be multiple; for even from birth, when they have been mutilated in the regions [of the body] useful for intercourse, both women and men are born sterile, so that the former don't reach puberty and the latter don't grow beards but continue to be like eunuchs; whereas to some it happens
25 that they suffer the same [thing] as they advance in age, sometimes on account of their bodies being well fed (for in the [women] who

---

61 Cf. *On the History of Animals* 566a27–31.

become too fat and in the [men] in too good a condition, the spermatic residue is used up for their body, and the former don't have menses and the latter [don't have] semen), while at other times on account of disease the [men] emit [seed] that is wet and cold, and   30
in the women the [menstrual] purgations [are] inferior and full of morbid residues. And in many [men] and many [women] this condition comes about also on account of deformities around the parts and the regions that are useful in connection with intercourse. Some of such [conditions are] curable, but others [are] incurable, and [they] remain sterile especially if they have come into being[62]   35
of such a sort in their first constitution; for there are born mannish   747a
women and womanish men, and in the former the menses don't come to be, while in the latter the seed is thin and cold. Wherefore reasonably the seed of the men is examined [as to] whether it is sterile by means of tests in water; for what is thin and cold quickly   5
disperses on the surface, while what is fertile goes to the bottom; for what is concocted is hot, and [it is] that which is set and has thickness [that] has been concocted. And they examine the women with pessaries, [to see] if the odors reach from below up to the breath [that goes] outside, as well as with colors rubbed into their   10
eyes, [to see] if they color the spit in the mouth. For if these [things] don't happen, it makes clear with regard to their body that the channels through which the residue is secreted have been obliterated and grown together. For the place around the eyes, of those in the region of the head, is the most full of seed. And this is made clear because it alone noticeably changes form during acts of intercourse, and in those who engage in too many sexual acts, their   15
eyes manifestly sink in. Responsible [for this is] that the nature of semen is similar to that of the brain; for its material is watery, whereas its heat is a supplemental acquisition. And the spermatic purgations are from the diaphragm, for the ruling beginning of   20
the [animal's] nature is from there, so that the motions from the genitals reach the chest; and all the odors out of the chest produce sense perception through breathing.

---

62 Omitting, at 746b35, the definite article added without any ancient authority by Drossaart Lulofs, and which doesn't really affect the meaning.

Chapter 8

Now then in human beings and the other kinds, as was said earlier,
25   such deformation occurs in a portion [of each], but the kind [con-
sisting] of mules is wholly sterile. Concerning the cause, in the way
Empedocles and Democritus speak—the former speaking obscurely,
and Democritus more intelligibly—they have not spoken finely. For
they state the demonstration [as applying] equally to all those that
couple outside of their own kind. For Democritus, on the one hand,
30   asserts that the channels of the mules have been destroyed in the
uterus as a result of the ruling beginning of the animals not having
come into being from [parents] of the same kind. But it happens that
this [sort of origin] is found in the case of other animals, but that they
generate none the less, and yet if this were indeed responsible for the
sterility [of mules], the others that have intercourse in this manner
ought also to be sterile. Empedocles, on the other hand, alleges as
35   cause that the mixture of the seeds becomes dense, out of each of the
747b   two semens being soft; for [he says that] the hollows of the one fit together with the densities of the other and [that] out of such [things],
out of softs, there comes to be a hard, just as copper mixed with tin,
neither giving the cause correctly in the case of copper and tin (it has
5   been spoken in the *Problems* about them)[63] nor, in general, making
his ruling beginnings out of [things that are] intelligible. For how do
the hollows and the hardnesses fitting together, the ones to the others,
produce the mixture, for example, of wine and water? This statement
is beyond us; for the way in which we have to take the hollows of wine
10   and of water is exceedingly contrary to sense perception. Further,
since it happens also that from horses a horse comes into being and
from asses an ass and from a horse and an ass a mule, in both ways,
whichever of the two is male and [whichever] female, why from these
does there come into being [seed] so dense that what is born is sterile,
15   but from a female and a male horse or a female and a male ass it isn't
born sterile? And yet both the [seed] of the male horse and that of the
female [are] soft, and both the female horse and the male have intercourse with the ass, both [the one] with the male and [the other] with
the female. And as a result of this, sterile [offspring] come into being

---

63 There is no such discussion in the extant portions of the *Problems*.

from both [of these couplings], as he asserts, because from both, from the [two] seeds that are soft [in each case], a single [dense seed] comes into being.[64] But then that which comes into being from a male and   20
a female horse ought also [to be the same]. For if only one of the two [seeds, that from the male horse or that from the female,] mixed [with that of an ass so as to produce an offspring], it would be possible to say that that one [is] responsible for [the mule's] not generating, [being] similar[65] to the semen of the ass. But as things stand, of whatever sort that [semen from the ass] is with which it mixes, the [semen with which it mixes] from the [animal] of the same kind [is] also of that sort. Further, the demonstration has been spoken as applying equally to both, both the female [mule] and the male, whereas the   25
male does generate [from a mare], as [people] assert, when he is seven years old only, but the female is wholly sterile, [and] even she because of her not bringing [the offspring] to birth, since a mule has in fact carried an embryo.[66] But perhaps a more convincing demonstration than those that have been mentioned might seem to be one based on reasoning. I say "based on reasoning" on account of this, that insofar as it is more general, it is further from the ruling beginnings peculiar   30

---

64 At 747b19, Drossaart Lulofs would add the word *puknon* at the end of this phrase, although there is no ancient authority for the addition. On the basis of his text, the word "dense," which I have included here within square brackets, would be part of the translation proper.

65 At 747b22, I read *homoion*, the reading of the preponderance of the manuscripts, rather than Drossaart Lulofs' reading of *oukh homoion on*, whose ancient support is from only one manuscript, and from the Arabic translation and the Latin translation by William of Moerbeke. Drossaart Lulofs' reading would require, instead of the translation "[being] similar . . . ," the translation "because of its not being similar . . . ." I interpret the similarity that Aristotle is speaking of on my reading as similarity in softness, such that when the two seeds are mixed together, there comes to be, according to Empedocles, a single dense seed that gives rise to an infertile mule. And if only one sex of horse could breed with an ass, Empedocles could say that the seed from the horse of the other sex would not have a similar softness, and so the offspring of the two horses would not have to be sterile.

66 In *On the History of Animals* 577b19–23, Aristotle speaks both of a female mule having become pregnant but failing to bring the embryo to term, and of a male mule having generated an offspring from a female horse. He also says there that it is only at the age of seven, not earlier and not later, that male mules have intercourse. Regarding a male mule generating an offspring from a female horse, see also 748b31–34 below.

[to the particular subject]. It is somewhat as follows: if from a male and a female of the same species there naturally comes into being a male or a female of the same species as those that generated it, as for example, from a male and a female dog a male or a female dog, and from those different in species [there comes into being an offspring] different in species, for example, if a dog is different from a lion, and from a male dog and a female lion [there comes into being something]
35   different, and from a male lion and a female dog [something] differ-
748a   ent; so that[67] since a male and a female mule come into being, with them not being different in species from one another, and a mule comes into being from a horse and an ass, and these and mules are different in species, [it is] impossible [for anything] to come into being from mules; for a different kind [is] not possible, on account of there coming into being from a male and a female, those of the same
5   species, [an offspring] the same in species, and a mule [is not possible] because it comes into being from a horse and an ass, which are different in species, and it was posited that from those that are different in species a different animal comes into being. But this argument is too general and empty. For arguments [that are] not from the ruling beginnings peculiar [to a particular subject are] empty, but seem to be about the things [that really are], though they are not. For [it is]
10   the [arguments] from the geometric ruling beginnings [that are] geometric, and likewise also in the case of the others; but what is empty, while it seems to be something, is nothing. And [the argument in question is] not true, because many of those coming into being from [parents] not of the same species[68] come into being fertile, as was said earlier. So then one ought not to inquire in this manner, either concerning the other [things] nor the natural [ones]. One would grasp
15   the cause more by looking into it on the basis of the [characteristics] that belong to the class of horses and that of asses, that, first, each of the two of them gives birth to one offspring at a time from the animals of the same kind, [and,] next, the females are not always able to con-

---

67 The translation "so that" is a faithful rendering of the Greek conjunction *hōste*, which, like the English, seems to disregard the fact that the premises of this conclusion were introduced in an if-clause.

68 At 748a12, Drossaart Lulofs would insert the preposition *ex*, an addition that makes the Greek construction easier than what appears in the manuscripts, but doesn't affect the translation.

ceive from the males, wherefore [breeders] put the [male] horses [to the females] at intervals, because of their not being continuously able to bear.[69] Rather, the mare is not prone to menstruate, but discharges the least [menstrual blood] of the quadrupeds. And the female ass doesn't [readily] accept impregnation, but expels the semen along with its urine, wherefore they follow after [it] and flog [it].[70] And further, the animal, the ass, is cold, wherefore, because of their nature being intolerant of cold, they don't usually come into being in wintry places, for example, around Scythia and the neighboring region, nor in the vicinity of the Celts beyond Iberia; for this region [is] also cold. And also on account of this cause, they put the males to the [female] asses not, as with horses, at the equinox, but around the summer solstice, so that the foals might be born in the hot season; for they are born in the same [season] in which they are mated; for both horse and ass are pregnant for a year. And since [the ass] is cold, as was said, in its nature, [it is] necessary also for the semen of such [an animal] to be cold. And [here is] a sign of this: on account of this, if a horse mounts a [mare] that has been impregnated by an ass, it doesn't destroy the impregnation from the ass, but if the ass mounts [a mare that has been impregnated by a horse], it destroys that from the horse on account of the coldness of its seed. And so when the seeds are mixed with one another, [the impregnation] is saved on account of the heat of the one; for the [seed] secreted from the horse is hotter; for both the material and the semen of the ass [are] cold, whereas the [semen] of the horse [is] hotter. And when either a hot [seed] is mixed with a cold [one] or a cold [one] with a hot [one], it results that the embryo itself that comes into being from these is saved, i.e., these [animals] are fertile from one another, but the [offspring] from these [is] no longer fertile, but sterile as regards the generating of perfected offspring. And in general, with each of the two already being naturally disposed to sterility, for the other [things] that have been mentioned belong to the ass, and [also] if it doesn't begin to generate after the first shedding [of teeth], it no longer generates at all; so close is the

20

25

30

35
748b

5

10

---

69 At 748a19–20, Drossaart Lulofs would delete the words translated as "because of their not being continuously able to bear," though they appear in all the manuscripts.

70 Cf. *On the History of Animals* 577a21–24, where Aristotle makes it explicit that they follow it in order to keep it from urinating.

body of asses to being sterile.[71] And likewise also the horse; for [it is] naturally disposed to sterility, and [only] this much is lacking for it to be sterile, that the [seed produced] from it become colder; and this comes about when [its seed] is mixed with the secretion from the ass.

15 And the ass comes so close to generating a sterile [offspring] in its coupling with its own [kind] that when what is contrary to nature is added, if in that case, from one another, they were only with difficulty able to generate one [offspring], the [offspring] from these, [being] still more sterile and contrary to nature, will lack nothing that belongs to being sterile, but will be sterile from necessity. And it happens also

20 that the bodies of mules become big on account of the secretion [that would go] to the menses being diverted toward growth. And since the pregnancy of such [animals lasts] for a year, the [female] mule must not only conceive but nourish completely; and this [is] impossible if the menses don't come about. And they don't come about in [female]

25 mules, but their useless [nutriment] is excreted along with the residue from the bladder (wherefore also the male mules don't sniff at the genitals of the females, as the other solid-hoofed [animals] do, but at the residue itself), whereas the rest [of the nutriment] is diverted toward the growth of the body and its size. So that it is sometimes poss-

30 ible for the female to conceive, which has manifestly already happened, but [it is] impossible [for it] to complete the process of nourishing [the embryo] and to bring it to term. And the male might sometimes generate, on account of the male being hotter than the female by nature, and on account of the male not contributing any body to the mixture. What is brought to perfection becomes a *ginnos*.

35 This is a deformed mule; for *ginnoi* come into being also from the [mating of the] horse and the ass, when the embryo has become dis-

749a eased in the uterus. For the *ginnos* is like the *metakhoira* among pigs;[72] for there too what has been deformed in the uterus is called a *metakhoiron*, and any chance one among the pigs may come into being of that sort. And also dwarfs come into being in the same way;

5 for these too become deformed with respect to their parts and their size during gestation, and they are like *metakhoira* and *ginnoi*.

---

71   This sentence, if it is a sentence, lacks a main clause also in the original Greek.

72   The Greek word for pig is *khoiros*, and the word *metakhoira*, which I don't attempt to translate, has sometimes been translated as "after-pigs."

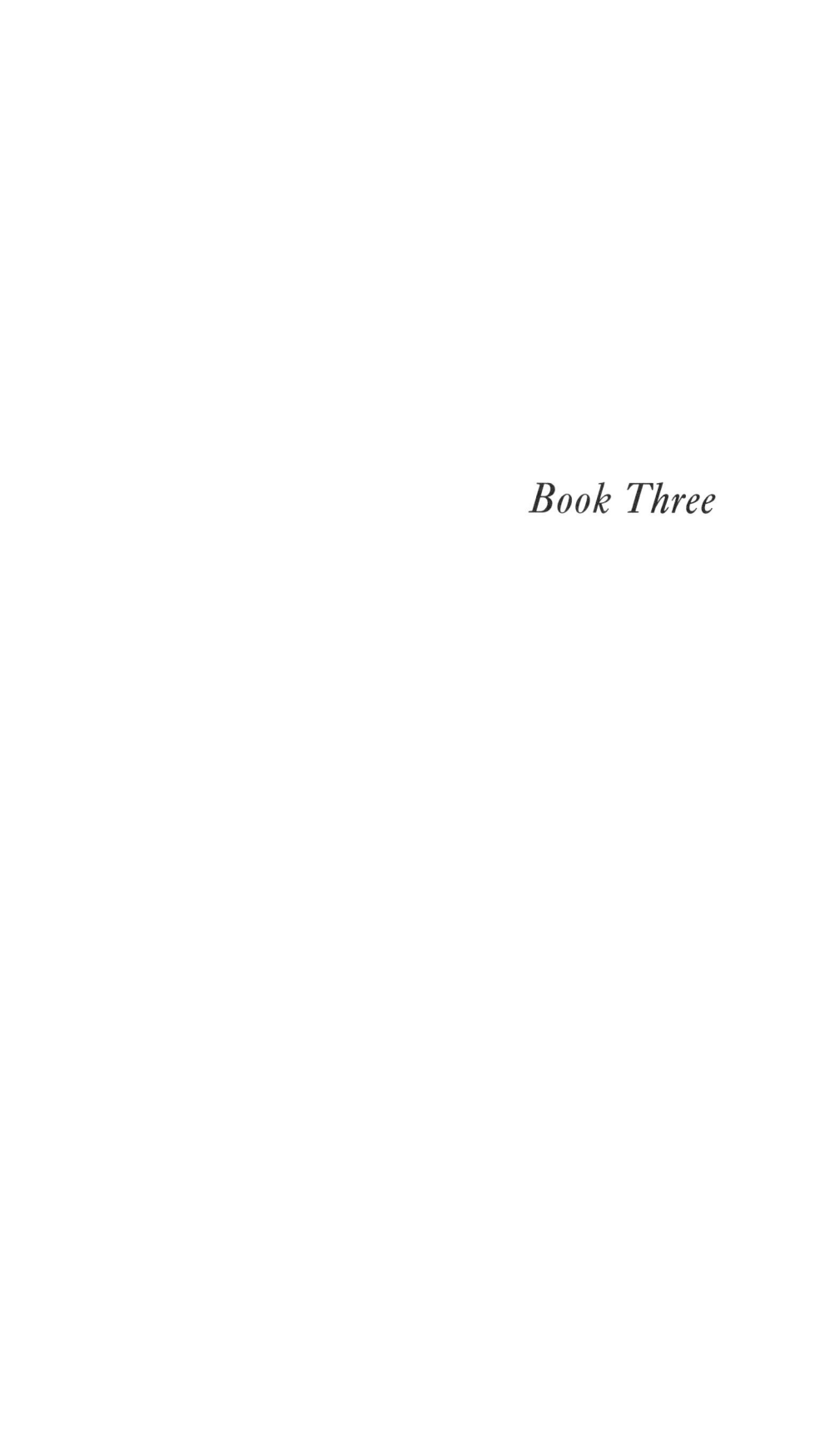

Book Three

Chapter 1

And so about the barrenness of mules it has been spoken, and about   10
the [animals] that bring forth animals both externally and inside
themselves; but in those of the blooded [animals] that lay eggs, in
one way the [things] having to do with their comings into being are
about the same as in the case of the [animals] that walk,[1] and it is
possible to grasp something [that is] the same concerning them all,
but in another way they have differences both in relation to each
other and in relation to those of the animals that walk. Now in gen-   15
eral all [blooded animals] come into being from coupling and from
the male emitting semen into the female; and of those that lay eggs,
the birds release a perfected and hard-shelled egg, if it hasn't been
deformed somehow on account of sickness, and the [eggs] of birds
are all two-colored, whereas among the fish, the selachian-like, as   20
has often been said, bring forth animals after laying eggs inside
themselves, with the egg having moved from one region of the
uterus to another, and their egg is soft-shelled and uniform in color.
One alone of such [fish], the so-called fishing-frog, doesn't bring
forth animals inside itself; concerning which the cause must be
stated later. The others of the fish, as many as lay eggs, release an   25
egg that has one color, but this [is] unperfected; for it obtains growth
externally due to the same cause as that due to which those of the
eggs that are perfected inside [do]. Now concerning uteruses, what
differences they have, and on account of what causes, has been stated

---

1 By distinguishing here the blooded animals that lay eggs from those that
   walk, Aristotle seems to ignore those egg-layers that are also walkers, even
   though he has said earlier that there are many of these, for instance, lizards
   (732b15–24), and will speak explicitly of them again in the next chapter
   (752b35–753a7; 754a15–18). I am unable to explain why he does this.

before.[2] And indeed, among those that bring forth animals some
30 have the uterus above by the diaphragm, whereas others [have it]
below by the genitals; those that are selachian-like [have it] above,
whereas those that bring forth animals both inside themselves and
outside, for example, human being and horse and each of the other
such [animals], [have it] below. And of those that lay eggs, some
[have it] below, like those of the fish that lay eggs, whereas others
35 [have it] above, like the birds. Now embryos in birds are also con-
749b stituted spontaneously, which some call wind eggs and west wind
eggs, and these come into being in those of the birds that are neither strong flyers nor crook-taloned, but in the prolific, because
these have much residue (whereas in those that are crook-taloned,
5   the secretion of this sort is diverted to their wings and their feathers and they have a body [that is] small and [also] dry and hot),
and menstrual secretion and semen are residue; since, then, both
the nature of wings and that of seed come into being out of residue,
nature is not able to produce much [of it] for both. And also on ac-
10   count of this same cause, those that are crook-taloned neither copulate frequently nor are they prolific, whereas those that are heavy,
and all those of the strong flyers whose bodies are bulky, like [the]
pigeon and those of that sort, [are prolific]. For in those that are
heavy and not strong flyers, for example, chickens and partridges
and the others of that sort, there comes to be much of that sort of
15   residue; wherefore their males copulate frequently and their females emit much material, and some of such [birds] lay many
[eggs] and others [lay them] frequently, for example, a chicken and
a partridge and the Libyan ostrich [lay] many, whereas those of the
pigeon family don't [lay] many, but [they lay them] frequently; for
these are intermediate between the crook-taloned and those that
20   are heavy; for they are strong flyers like the crook-taloned, but they
have bulky bodies like those that are heavy, so that because of their
being strong flyers and their residue being diverted to that, they
lay few, but because of the size of their body and because of their
having a stomach that is hot and very much able to concoct, and
25   in addition, because of their obtaining food easily, whereas the
crook-taloned [do so] with difficulty, [they lay] frequently. Also
those of the birds that are small copulate frequently and are pro-

---

2   716b32–717a10; 718a35–b27.

104

lific, just as sometimes also among the plants; for the [material for] growth in the body becomes spermatic residue. Wherefore also, among hens, the Adriatic ones lay the most [eggs]; for because of the smallness of their body, their nutriment is expended on pro-   30 creation. And the lowbred lay more [eggs] than the highbred; for their bodies are wetter and bulkier, whereas those of the others are leaner and drier; the highbred spirit comes about more in bodies of this [latter] sort. Further, the thinness and weakness of their legs also contributes to the nature of such [birds] being given to   35 frequent copulation and [being] prolific, as also in the case of   750a human beings; for in those of such a sort, the nutriment for the limbs is diverted to spermatic residue; for what nature takes away from there, it adds here. Those that are crook-taloned, on the other hand, have a strong foot and legs that have thickness, on account   5 of their manner of life; so that on account of all these causes, they neither copulate frequently nor are they prolific. The kestrel is the most prolific [of them]; for pretty much alone among those that are crook-taloned, it drinks, and its wetness, both that which is innate and that which is taken in, together with the heat that is present in it, [is] productive of seed. But not even this one lays very   10 many [eggs], but four at most. The cuckoo lays few [eggs], even though it is not crook-taloned, because it is cold in its nature (the cowardice of the bird makes [this] clear), whereas an animal that is productive of [much] seed has to be hot and wet. And [it is] manifest that [it is] cowardly; for it is chased away by all birds and it lays [eggs] in others' nests. Those of the pigeon family are accus-   15 tomed to lay two [eggs] for the most part; for they don't lay just a single [egg] (for no bird lays just a single [egg] except the cuckoo, and this one sometimes lays two), nor do they lay many, but they produce [them] often, two or three at most, but for the most part two; for these numbers are in between one and many. That the nu-   20 triment in those that are prolific is diverted to seed [is] manifest from what (pl.) happens. For the majority of trees that have produced too much fruit dry out after their bringing forth, whenever nutriment isn't left over for their body, and annuals seem to suffer the same [thing], for example, leguminous plants and corn and the   25 others of that sort; for they use up all their nutriment on seed; for their kind is productive of much seed. And some hens, having laid too many [eggs], to the extent of even laying two in a day, have died

after laying [so] many. For these become violently purged as it were, both the birds and the plants; and this condition is an excess of ex-
30   cretion of residue. And such a condition [is] responsible also for the later sterility in the lion; for at first it brings forth five or six, then in the next year four, and again three cubs, then the next num-ber, [continuing] down to one, then none, suggesting that its resi-
35   due is being used up and that its seed is diminishing together with
750b   its prime of life coming to an end. And so in which of the birds there come to be wind eggs, and further, which sorts of them are prolific and unprolific and on account of what causes, has been said. Wind eggs come into being, as has been said earlier, because
5   of there being present in the female material that is spermatic and there not coming to be in birds secretion of the menses, as in the blooded animal-bearing [animals]; for in all these it does come to be, in some more, in others less, and in some [only] of such an amount as to give an indication. And likewise, neither [does it come to be] in fish, just as [it doesn't] in birds; wherefore also in these
10   there comes to be, without copulation, a constituting of embryos, in the same way as in birds, but less conspicuously; for their nature is colder. The secretion of menses that comes to be in the animal-bearing [animals] acquires consistency [instead] in birds at the times proper for the residue, and on account of the region by the
15   diaphragm being hot, [these eggs] are perfected in size, but both these and likewise those of the fish [are] imperfect in regard to gen-eration [*genesin*] without the semen of the male. The cause of these [things] has been stated earlier.[3] Wind eggs don't come into being in those of the birds that are strong flyers, on account of the same cause as that due to which such [animals] also don't lay many [eggs]; for in the crook-taloned the residue is small [in amount],
20   and they need the male in addition to give the impulse for the dis-charge of the residue. The wind eggs [that do] come into being [are] more [in number] than those [eggs] that come into being by procreation,[4] but smaller in size, on account of one and the same

---

3   Cf. 737a27–34; and also 741a6–32, although at the end of the latter passage, Aristotle says that the cause in question will be determined later on, thus im-plying that it hasn't yet been stated.

4   At 750b21–22, I read *tōn gonōi gignomenōn*, with Z, rather than *tōn gonimōn ōōn* with Drossaart Lulofs, who follows only one relatively late manuscript

cause; for on account of their being unperfected [they are] smaller in size, and on account of their size [being] smaller [they are] more in number. And [they are] also less pleasant [to eat] on account of   25
being more unconcocted; for in all [things] what is concocted is sweeter. Now that neither the [eggs] of birds nor those of fish are perfected in regard to the coming into being [of offspring] without males has been sufficiently observed, but as for embryos coming into being also in fish without males, [this has] not [been seen] to the same degree, but the occurrence has been seen[5] above all in the   30
case of river fish; for some manifestly have eggs straightaway, as has been written about them in the *Histories*.[6] But in general, in birds at any rate, even the eggs that come into being as a result of copulation are unable for the most part to obtain growth if the [female] bird is not continually mated. Responsible for this [is] that, as in the case of women to have sexual relations with males pulls   35
down the secretion of menstrual fluids (for the uterus, after having   751a
been heated, draws in the fluid and the channels are opened up), this happens also in the case of birds, with the menstrual residue advancing little by little, which [residue] is not secreted externally, on account of there being little [of it] and the uterus [being] up by   5
the diaphragm, but trickles together into the uterus itself. For this, that which flows through the uterus, causes the egg to grow, just as [that which flows] through the umbilical cord [causes] the em-

---

(though with support from the Arabic translation). According to Drossaart Lulofs' text, instead of "than those [eggs] that come into being by procreation," the translation would read "than the fertile eggs."

5   At 750b30, I read *to sumbainon*, with Z, rather than *sumbainon* with Drossaart Lulofs, a reading that has no ancient authority. According to Drossaart Lulofs' text, the beginning of this clause would be translated, instead of as "but the occurrence has been seen," as "but it has been seen occurring." More importantly, however, I agree with Drossaart Lulofs not to include the words *peri tous eruthrinous* as part of this clause, even though they appear in almost all the manuscripts, and are further supported by the ancient translations. These words would specify the erythrinus (cf. 741a36) as the river fish in which, above all, embryos have been seen coming into being without a male contribution. But a reference to this fish is not supported by the passage from *On the History of Animals* to which Aristotle alludes later in the sentence, and moreover, he considered this to be a deep-sea fish, not a river fish (cf. *On the History of Animals* 598a13).

6   *On the History of Animals* 567a28–32.

bryos (*embrua*) of the animal-bearing [animals to grow], since when birds have been mated once, they all continue pretty much always having eggs, but very small [ones]. Wherefore also concerning wind
10   eggs, some people are accustomed to speak [of them] as not coming into being, but as being leftovers from an earlier act of copulation. But this is false; for there have been sufficiently observed, both in the case of young chickens and that of goslings, wind eggs that come into being without copulation. Further, female partridges,
15   the unmated as well as the mated ones of those that are led out to the hunts [as decoys], when they smell the male and hear his call, the ones become full [of eggs] and the others lay [theirs] at once. Responsible for this happening [is] the same [thing] as in the case of human beings and quadrupeds; for if their bodies happen to be aroused for intercourse, some emit seed [merely] from seeing and
20   others when a slight touch has occurred. And birds of this sort are, in their nature, given to frequent copulation and abundant with seed, so that they need [only] a small motion when they happen to be aroused and the secretion [of their seed] comes about quickly, so that wind eggs are constituted in the unmated, while in those that have been mated [their eggs] grow and are perfected quickly.
25   Of those that lay eggs externally, birds release their egg [already] perfected, whereas fish [release theirs] unperfected, but it obtains growth outside, as has been said earlier. Responsible [for this is] that the genus of fish is prolific; it is impossible, then, for many [eggs] to attain perfection inside, for which reason they lay them
30   outside. And the emission is quick; for the uterus of fish that lay eggs externally is by the genitals. The [eggs] of birds are two-colored, whereas those of all fish [are] single-colored. And one might see the cause of two-coloredness from the capacity of each of the two parts, the white and the yellow. For the secretion [of the egg] comes about out of the blood (for no bloodless animal lays eggs),[7]
751b   and that blood is the material for bodies has often been said. Now the one [element] of it, the hot, is closer to the shape of the [off-

---

7   Some modern scholars think that the words translated as "for no bloodless animal lays eggs" should be deleted, even though they appear in all the manuscripts,
on the grounds that Aristotle has said earlier, at 733a26–29 and 733b7–10, that
soft-shelled animals and cephalopods, which are bloodless (cf. 720b2–5), lay eggs.
I am inclined to think that this clause may well be a scribal addition, but I translate it here out of deference to the manuscript tradition.

spring] that are coming into being; whereas what is earthier provides the consistency of the body and is further removed. Wherefore in as many eggs as are two-colored, the animal gets the ruling   5
beginning of its coming into being out of the white (for the soul-
ish ruling beginning is in what is hot), but its nourishment out of
the yellow. Now in those of the animals that are hotter in their nature, that out of which the ruling beginning comes to be and that
out of which they are nourished are set apart, and the former is
white and the latter yellow, and [there is] always more of what is   10
white and pure than of what is yellow and earthy. But in the [ani-
mals] that are less hot and more wet, [there is] more of what is yellow and [it is] wetter. Which happens in the case of marsh birds;
for they are wetter in their nature and colder than land birds, so
that also the eggs of such [birds] have much of the so-called yolk
and [it is] less yellow on account of the white being less separated.   15
And when we come to those of the egg-laying [animals] that are
actually cold in their nature and still wetter (and of this sort is the
class of fish), [their eggs] don't even have the white set apart on
account of their smallness and on account of the large amount of
what is cold and earthy [in them]; wherefore all the [eggs] of fish
are single-colored, and white as compared with yellow and yellow   20
as compared with white. But the [eggs] of birds, even their wind
eggs, have this two-coloredness; for they have that from which each
of the two parts will be—both that from which the ruling beginning [comes] and that from which the nourishment [comes]—but
these are unperfected and they need the male in addition; for wind
eggs become fertile if within a certain period of time [the birds]   25
are mated by the male. But [what is] responsible for their two-coloredness is not the male and the female—in the sense that the
white is from the male and the yellow from the female—but both
come into being from the female, only the one [is] cold and the
other hot. So then in as many [eggs] as contain much of what is
hot, it is separated off, whereas in as many [as contain] little [of it],
it is not able [to be separated]; wherefore the embryos of such [ani-   30
mals are] single-colored, as has been said. The semen has merely
caused [them] to set; and on account of this the embryo in birds
appears white and small at first, but as it progresses [it becomes]
all yellow, as more that is bloodlike is continually being mixed in;
and finally, with what is hot being separated off, the white encom-   752a

passes [the rest] all around, just as when a liquid boils evenly everywhere; for the white is by nature wet, but it has the soul-ish heat
within itself; wherefore it is separated off all around, and what is
yellow and earthy [is enclosed] inside. And if someone pours many
5 eggs together into a bladder or something like it, and boils [them]
in a fire that doesn't cause the motion of the heat to be faster than
the separation in the eggs, just as in one egg, also in the conglom-
eration of all the eggs, the yellow comes to be in the middle and
the white all around [it]. And so it has been said why some eggs are
10 single-colored and others two-colored.

Chapter 2

In the eggs, the ruling beginning of the male becomes separated
off at the place where the egg is attached to the uterus, and so the
egg [if it is one] of the two-colored ones[8] becomes non-uniform
and not altogether round, but more pointed at one end, on account
of [the part] of the white in which it contains the ruling beginning
needing to differ [from the rest]. Wherefore the egg is harder there
15 than below; for it has to shelter and protect the ruling beginning.
Also on account of this, the pointed [end] of the egg comes out
later; for what is attached [to the uterus] comes out later, and [the
egg] is attached near the ruling beginning, and the ruling beginning [is] in the pointed [end]. And it is the same way also in the
seeds of plants; for in some the ruling beginning of the seed is at-
20 tached [to the plant] in the branches, in others in the husks, and
in others in the seed cases. And [this is] clear in the case of the
leguminous [plants]; for where the two valves of the beans and of
the seeds of this sort are joined together, that is where they are attached [to the plant]; and the ruling beginning of the seed [is]
there. Someone might be perplexed about the growth of the eggs,
25 in what manner it comes from the uterus. For [unborn] animals
obtain their nourishment through the umbilical cord, but through

---

8 At 752a12, I read *to tōn dikhroōn öion*, with the great preponderance of the
manuscripts, rather than Drossaart Lulofs' *to dikhroon öion*, which he reads with
apparent support only from the Arabic translation. According to Drossaart
Lulofs' text, the translation would read "the two-colored egg" instead of "the
egg [if it is one] of the two-colored ones."

what do the eggs [obtain theirs]? For they don't, like larvae, obtain
their growth themselves through [or by] themselves. And if there
is something through which they are attached [to the uterus],
where does this go when [the egg] has reached perfection? For it
doesn't come out with it, as the umbilical cord [does] in the case
of [newborn] animals; for when it has reached perfection its sur-   30
rounding [envelope] becomes a shell. Now what has been said is
correctly asked about; but it escapes notice that the shell that
comes into being is at first a soft membrane, and it becomes hard
and brittle [only] when [the egg] has reached perfection, in such a
suitable way that it comes out still soft (for [otherwise] it would
produce suffering as it was being laid), but upon coming out it im-   35
mediately solidifies from having been cooled, with its wetness
evaporating quickly because of its small amount, and what is earthy   752b
being left behind. Now something of this membrane is to begin
with like an umbilical cord at the pointed [end of the eggs] and it
sticks out, while they are still small, like a pipe. [This] is manifest
in the abortions of small eggs; for if the bird aborts, from having
been drenched or chilled in some other way, the embryo appears   5
still bloodlike and manifestly has a small band, like an umbilical
cord, [going] through it. But as [the egg] becomes bigger, this is
twisted around more and becomes smaller. And when [the egg] has
reached perfection, this turns out at last [to be] the pointed [end]
of the egg. And underneath this [is] the inner membrane, which
separates the white and the yellow from it. And when it has reached   10
perfection, the whole egg is released and the umbilical cord doesn't
appear, as is reasonable; for it is the tip of its [pointed] extremity.
Delivery comes about for eggs in the opposite way than [it does]
for those that are born as animals; for these latter, namely, [it is]
with the head and the ruling beginning first, but for the egg, delivery comes about feet first, as it were. Responsible for this [is]   15
what has been said, that [the egg] is attached at its ruling beginning. The coming into being out of the egg takes place for birds
from the [mother] bird sitting on [it] and helping to concoct [it],
the animal being separated off out of a part of the egg, but obtaining its growth and coming to perfection out of the remaining part;
for nature puts the material of the animal in the egg together with   20
sufficient nutriment for its growth; for since the [mother] bird is
unable to perfect [the animal] inside herself, she lays the nutriment

111

in the egg along with [it]. For in the case of those born as animals, the nutriment, that which is called milk, comes into being in another part, in the breasts; but in the case of birds, nature produces

25   this in the eggs, in the opposite way, however, from what human beings suppose and Alcmaeon of Krotona asserts. For [it is] not the white [that] is milk, but the yellow. For this is the nutriment for the chicks, but they suppose [it is] the white on account of the similarity of color. Now then the chick comes into being from the [mother] bird sitting on [the egg], as has been said. Nevertheless,

30   if the season [is] temperate or the place in which [the eggs] happen to lie [is] sunny, both the [eggs] of birds and those of the egg-laying quadrupeds also become fully concocted [without incubation]; for [the quadrupeds] all lay [their eggs] on the earth, and they are concocted by the heat in the earth. And as many of the egg-laying

35   quadrupeds as visit [their eggs] and sit on [them], these do so more
753a   for the sake of protection. The eggs of birds and those of the quadruped animals come into being in the same manner; for [these] too are hard-shelled and two-colored, and they are constituted near the diaphragm, just like those of birds, and all the other [things] turn out the same both inside and outside [them], so that

5   there is the same inquiry concerning the cause of [them] all. But the [eggs] of the quadrupeds, on account of their strength, become fully concocted by the agency of the season, whereas those of the birds [are] more delicate, and they need the [mother] that laid [them]. And it looks even as if nature wishes to produce attentively caring perception directed toward the young;[9] but in the inferior

10   [animals] she imparts this only up until the [mother's] giving birth, in others also with regard to the coming to perfection [of the young animals], and in as many as are more intelligent, also with regard to their rearing. And indeed in those that have the greatest share of intelligence, there comes to be intimacy and love even toward those that have reached their [full] perfection, as in human beings and some of the quadrupeds, but in birds [this continues] up until

---

9   At 753a8, I read *tōn teknōn aisthēsin*, the reading of two of the oldest manuscripts, instead of Drossaart Lulofs' *tēn tōn teknōn aisthēsin*, which appears in most of the others. According to Drossaart Lulofs' text, the first part of this sentence would be translated as "And it looks even as if nature wishes to produce the attentively caring perception directed toward the young [that exists in animals]."

their having generated and reared [their young]; wherefore also,   15
the female [birds] that don't sit on [their eggs] when they have laid
[them] are in an inferior condition, as if deprived of one of their
innate [endowments]. Animals are perfected inside the eggs more
quickly during the sunny days; for the season contributes to the
work; for indeed concoction is [a product] of heat.[10] For the earth
contributes to concoction by its heat and the [mother bird] sitting   20
on [the eggs] does this same [thing]; for she pours in additionally
the heat inside herself. And eggs are spoiled and what are called
ouria[11] come into being more during the hot season, reasonably;
for just as wines become vinegary during hot spells from their sediment getting stirred up (for this [is] responsible for their being
spoiled), so also the yolk in eggs [causes them to go bad]; for this   25
in both [of them is] the earthy [part], wherefore also the wine becomes turbid when its sediment gets mixed in, as do the eggs that
spoil when the yolk does. Now this sort [of thing] happens reasonably with those [birds] that lay many [eggs] (for [it is] not easy to
give the fitting amount of heating to them all, but [what happens
is that] it is too little for some, while for others it is too much and   30
makes them turbid, rotting [them] as it were), but with those that
are crook-taloned, though they lay few [eggs], this happens no less;
for often, even out of two, one or the other becomes rotten, and
the third [does so] always, so to speak; for being hot in their nature,
they make the liquid in their eggs boil over as it were. For indeed
the yellow and the white have [each] a nature that is opposite [to   35
the other's]. For the yellow congeals in frosts, but liquefies when   753b
it is heated; wherefore also, when it is concocted in the earth or by
being sat upon, it becomes liquid, and being of such a sort it becomes nutriment for the animals that are being constituted. But
when it is subject to fire and baked, it doesn't become hard, on account of its nature being earthy in the manner that wax is; and on   5
account of this, when [the eggs] are heated more [than the fitting

---

10   At 753a19, I read *thermotētos*, a modern emendation accepted by Drossaart
     Lulofs, rather than the manuscripts' *thermotēs tis*. According to the manuscript
     reading, the translation would be "concoction is a sort of heat" instead of
     "concoction is [a product] of heat," but that doesn't make sense to me.

11   The word *ourion* (the singular form of *ouria*) means primarily "of urine" or "like
     urine." Later in this chapter, at 753a32 and 753b7, I will translate it as "rotten."

amount], †if they are not†[12] [formed] out of a liquid residue, they turn urine-like and become rotten. But the white is not congealed by frosts, but is liquified rather (what is responsible [for this] has been said earlier),[13] but when subject to fire, it becomes solid;

10 wherefore also when it is concocted in connection with the coming into being of animals, it thickens. For the animal is constituted out of this, whereas the yellow becomes nutriment, and those of the parts that are being constituted at any given time have their growth from there. Wherefore the yellow and the white have been set apart by membranes, as having a nature that is different [each from the

15 other's]. Now [to see] with precision in what way these [parts] are related to one another, at the beginning of the process of coming into being as well as while the animals are being constituted, and further, concerning the membranes and the umbilical cords, one must contemplate on the basis of what has been written in the *Histories;*[14] but for the present inquiry [it is] sufficient that this much be manifest, that after the heart has been constituted first and the

20 great blood vessel has been marked off from it, two umbilical cords extend from the blood vessel, one to the membrane that surrounds the yellow and the other to the chorion-like[15] membrane that encloses the animal all around; and this latter one [goes] around [underneath] the membrane of the shell. Now through the one of them it gets its nutriment out of the yellow, and yet the yellow be-

25 comes larger; for it becomes more liquid as it is being heated. For the nutriment, though it is body-like, has to be a liquid, as [it does] for plants, and at first both the [animals] that come into being inside eggs and those [that do so] inside animals live the life of a plant; for they obtain their first growth and nutriment from some-

---

12 The daggers that I have included here, at 753b6, are from Drossaart Lulofs' text, and indicate that while he doesn't accept the manuscript reading, he can think of no emendation that he trusts sufficiently to include in his text. I don't know whether the manuscript reading is sound, but if it is, I think that by "liquid residue," Aristotle must mean a sufficient amount of liquid residue so that it doesn't "boil over."

13 This is perhaps a reference to 735a34–736a2, where, however, Aristotle is discussing the nature of semen.

14 *On the History of Animals* 561a3–562a21.

15 Cf. 739b31.

thing by being attached [to it]. The other umbilical cord extends to the surrounding chorion. For one has to take it that those of the animals that are born from eggs are related to the yellow, the yolk,[16] in the same way as fetuses (*embrua*) that are born as animals are related to the mother when they are inside the mother (for since the [animals] born from eggs are not nourished completely inside the mother, they take some part of her out [with them]), and [one has to take it that they are related] to the outermost membrane, the blood-like one, as [those others are] to the uterus. And at the same time, the analog of the uterus, the shell of the egg, naturally surrounds the yellow and the chorion, as if someone were to put [it] around the fetus (*embruon*) itself as well as around the whole mother. It is in this way because the fetus (*embruon*) has to be inside the uterus and in contact with the mother. Now in those that are born as animals the uterus is inside the mother, but in those that are born from eggs [it is] the reverse, as if one were to say that the mother is inside the uterus; for what comes to be from the mother, the nutriment, is the yellow. Responsible [for this is] that nourishing is not completed inside the mother. As the [fetuses] grow, the umbilical cord that goes to the chorion collapses before [the other one does], because here [is where] the animal has to come out, while the rest of the yellow and the umbilical cord that goes into the yellow [collapse] later; for the [animal] that has come into being has to have nutriment immediately; for it isn't nursed by its mother and it isn't immediately able to provide its nutriment through itself; for which reason the yellow goes inside [it] together with the umbilical cord, and the flesh grows around it. So then the [animals] that come into being externally out of perfected eggs come into being in this manner, in the case of birds as well as that of as many quadrupeds as lay an egg with a hard shell. These [things] are seen more clearly in the case of the larger [animals], for in the smaller ones, they are less apparent, because of the smallness of their bulks.

30

35

754a

5

10

15

20

---

16 At 753b31–32, Drossaart Lulofs would delete the words *ton neotton*, translated as "the yolk," even though they are present in all the manuscripts. The usual meaning of the word *neotton* is "chick," which of course doesn't make sense here, but Aristotle uses it once in *On the History of Animals* (565a3) to mean "yolk." And so even though these words might be a scribal interpolation, I have chosen to include them.

Chapter 3

In addition, the class of fish is egg-laying. And of these, those that have their uterus below lay an unperfected egg on account of the cause that has been stated,[17] whereas those of the fish that are called selachians lay a perfected egg inside themselves, but bring forth

25   animals externally, except for one, which they call a fishing-frog; this one alone lays a perfected egg externally. [A] cause [is] the nature of its body; for its head is many times bigger than the rest of its body, and this [is] spiny and extremely rough. Wherefore neither does it admit its young [into its mouth] afterward nor does it bring

30   [them] forth [as] animals from the beginning; for the size and the roughness of its head, just as it prevents [them] from entering in, so [would it] also from coming out. And since the egg of the selachians is soft-shelled (for they are unable to harden and dry its surrounding [envelope], for they are colder than birds), the egg of the fishing-frogs alone is solid and firm, for its preservation outside,

35   whereas those of the others [are] wet and soft in their nature; for
754b   they are sheltered inside the body of the [mother] that is carrying [them]. The coming into being out of the egg [is] the same for the fishing-frogs, which are perfected externally, and for those [selachians that are perfected] inside [the mother], and between these [latter] and the [eggs] of birds [it] is in a way similar, but in a way

5   different. For in the first place, [these] don't have the other umbilical cord, the one extending to the chorion that is underneath the surrounding shell; and responsible [for this is] that they don't have the surrounding shell; for [it would be] of no use to them; for their mother shelters [them], whereas for the eggs that are laid, the shell is protection against injuries from outside. In the next place, coming

---

17 Other translators treat this as a reference to Book One, Chapter 8, 718b21–24, even though Aristotle's explicit claim there—that the animals that lay unperfected eggs necessarily have their uterus below—is the converse of what he says here. However, Aristotle also argues in that chapter that the region in which eggs are perfected, in those animals that lay perfected eggs, is necessarily hot, and that heat is present in the upper region by the diaphragm, which is why it is necessary for these animals to have their uterus there. And this suggests the argument that those fish that have their uterus below necessarily lay unperfected eggs because their uterus isn't hot enough to be able to perfect them (718b15–21; cf. 733b8–10).

into being is from a tip of the egg for these as well [as for birds],   10
but not [from the one] where it is attached to the uterus; for birds
come into being from the pointed [end], and that is where the attachment of the egg was.[18] Responsible [for this difference is] that
the [egg] of birds comes to be separated from the uterus, whereas
the egg of [fish] of this sort, not of all but of most of them, is attached to the uterus [even] when perfected. And as the animal   15
comes into being at the tip, the egg gets used up (just as also in the
case of birds and the other [eggs] that have been detached), and in
the end the umbilical cord of the [animals] that are now perfected
is [still] attached to the uterus. It is similar[19] also for as many [sela-
chians] as whose eggs are detached from the uterus; for with some
of them, the egg is detached when it comes to be perfected.[20] Someone might be perplexed, then, as to why the comings into being of   20
birds differ in this way from those of [these] fish. Responsible [for
this is] that the [eggs] of birds have the white separated from the
yellow, whereas those of fish [are] single-colored, and that which is
of this sort [is] mixed throughout, so that nothing prevents [the
eggs] from having their ruling beginning at the opposite [end]; for
not only at the point of attachment is [the egg] of this sort, but also   25
at the directly opposite [end], and [it is] easier[21] [than if the white
and the yellow were separated] to draw nutriment from the uterus
by certain channels [extending] from this ruling beginning. [This
is] clear in the case of the eggs that are not detached; for in some of
the selachians the egg is not detached from the uterus but, being
connected to it, it descends toward its being brought forth as an
animal; in which [cases] the animal that has been perfected has its   30
umbilical cord from the uterus when the egg has been used up. [It
is] manifest, then, that even earlier, when the egg was still surrounding it, the channels extended to the uterus. And this happens, as we
said,[22] in the smooth dogfish. So then the coming into being of

18 Cf. 752a16–18.

19 That is to say, the animal comes into being at the non-attached tip of the egg.

20 Not only the fishing-frog. Cf. 737b18–21.

21 At 754b26, some manuscripts, though not the oldest ones, read *radion*, mean-
    ing "easy," instead of *raion*, which I have translated as "easier [than if the
    white and the yellow were separated]."

22 *On the History of Animals* 565b2–17.

[these] fish differs from that of birds in these respects and on ac-
35 count of the causes that have been stated. The other [things] occur
755a in the same manner. For they have the other umbilical cord in a like
manner, as birds [have it going] to the yellow, so [these] fish [have
it going] to the whole egg (for it doesn't have one [part] white and
the other yellow, but [it is] all of one color), and they are nourished
5 out of this, and as it is used up, the flesh encroaches and grows
around it in a similar way.

Chapter 4

So then concerning the [fish] that lay a perfected egg inside themselves and [then] bring forth animals externally, their coming into
being is in this manner, but most of the other fish lay eggs externally,
and all [of them] an unperfected egg except for the fishing-frog.
10 Concerning this one, what is responsible has been said earlier.[23] And
also concerning those that lay unperfected [eggs], what is responsible has been said.[24] The coming into being out of the egg, [that]
of these and [that] of the selachians that lay eggs inside [them-
selves], is in the same manner, except, indeed, that the growth [of
these is] fast and out of small [eggs] and that the extremity of the
15 egg [is] harder. The growth of the egg is similar to [that of] larvae;
for the animals that bring forth larvae also bring forth [something]
small at first, and this grows through [or by] itself and not through
any attachment. What is responsible [is] nearly the same as in the
case of yeast; for yeast too, out of [a] small [amount], becomes big,
as what is more solid becomes liquid and what is liquid becomes
20 *pneuma.* The nature of soul-ish heat fashions this in animals,
whereas in yeasts [it is] the heat of the juice that has been mixed
in. So then the eggs grow from necessity, on the one hand, on account of this cause (for they have a yeast-like residue), but on the
other hand, for the sake of what is better; for [it is] impossible for
25 them to obtain the whole of their growth in the uterus on account
of these animals laying many [eggs]. For on account of this they
are separated [from the mother while] extremely small and they

---

23 754a23–35.
24 718b5–10; and 751a24–29.

obtain growth quickly, small on account of the uterus being cramped in relation to the multitude of the eggs, and quickly so that their kind not be destroyed from their spending too much time, in their coming into being, with growth, since even now most of the em- 30 bryos that are laid are destroyed. Wherefore the kind consisting of fish is prolific; for nature fights against destruction by means of multitude. But there are some fish, for example, the so-called *belonē*, that burst on account of the size of their eggs;[25] for the embryos that this one carries, instead of being many, [are] big; for nature, having taken away from their multitude, has added to their size. So 35 then [the fact] that the eggs of this sort grow [after having been laid] and [the] cause on account of which [they do], has been stated. 755b

Chapter 5

That these fish too produce eggs [rather than larvae] a sign [is] that even those of the fish that bring forth animals, such as the sela-chians, first produce eggs inside themselves; for [it is] clear that the kind as a whole, that consisting of fish, is egg-producing. However none of such eggs, belonging to as many [kinds of fish] as are, the 5 one, female and, the other, male[26] and that come into being from copulation, attain perfection unless the male sprays its milt over [them]. But there are some who assert, speaking incorrectly, that all fish, except [among] the selachians, are females; for they suppose that the females differ from those of them held to be males just as [is the case] among plants, in as many [kinds] as in which the one 10 bears fruit and the other is without fruit, for example, olive and oleaster, and fig and caprifig;[27] [they suppose that this is the case] likewise with fish, except for the selachians; for they don't dispute about these. And yet the males, those that are selachian-like and those within the kind consisting of the egg-layers, have the same sort of structure with regard to their milt-producing [parts], and in season seed is manifestly squeezed out of both. Also, the females 15 have a uterus; but not only those that lay eggs but the others also

---

25 Cf. *On the History of Animals* 567b23–26.

26 Cf. 741a38–b2.

27 Cf. 715b21–25; and *On the History of Animals* 557b25–31.

ought to have [one], but differing from [that of] those that lay eggs, as do female mules within the kind consisting of the [animals] with long-haired tails,[28] if indeed the kind were all female but some of
20   them barren. But as it is, some have milt-producing [parts] and others a uterus, and in all [kinds] outside of two, erythrinus[29] and channa, there is this difference; for some have milt-producing [parts] and others a uterus. And the perplexity on account of which they take [things] in this way is easily resolved when [they] have heard what occurs. For they assert that none of the [animals] that copulate
25   bring forth many [offspring], speaking correctly, for as many animals as out of themselves generate either animals or eggs [that have been] perfected, [these] don't bring forth many on the scale that the egg-layers among the fish do; for the multitude of these eggs is something enormous. But they haven't considered this, that what (pl.) [happens] with regard to the eggs of fish isn't of the same sort as [is the case with] the [eggs] of birds. For birds and as many of the
30   quadrupeds as lay eggs, and if any of the selachian-like [fish do],[30] generate a perfected egg, and it doesn't obtain growth after having come out, but fish [lay their eggs] unperfected, and the eggs obtain growth outside. Further, it is the same way also in the case of the cephalopods and in that of the soft-shelled [animals], which are in-
35   deed seen coupling, on account of their coupling lasting for a long time. And of these, it is manifest that the one is male and that the
756a   other has a uterus. And [it is] strange also for this capacity[31] not to be present in every kind,[32] just as in those that bring forth animals one is male and the other female. Responsible for the ignorance of those who speak in that way [is] that the differences, which are of

---

28 At 755b18–19, Drossaart Lulofs would delete the words translated as "as do female mules within the kind consisting of the [animals] with long-haired tails," though without any manuscript authority.

29 Cf. 741a32–38.

30 Cf. 754a23–26.

31 By "this capacity," Aristotle means the capacity to copulate, male with female.

32 At 755b36, the Greek phrase that I have translated as "in every kind" is *en panti genei*, which appears in all the manuscripts and which presumably means "in every kind of fish." Drossaart Lulofs, however, would emend this phrase to read *en panti tōi genei*, which means "in all of the kind," i.e., "in all of the kind consisting of fish."

all sorts, concerning the copulations and the comings into being of animals are not clear, but that, theorizing from a few [cases], they suppose it ought to be similar in all. Wherefore also those who say that pregnancies result from the female fish gulping the [male] seed down speak in this way because of not having noticed some [things]. For at about the same time the males have their milt and the females their eggs, and the closer the female is to laying, the more [abundant] and the more liquid does the milt become in the male. And just as the growth of the milt in the male and that of the egg in the female happen at the same time, so also does their release; for neither do the females lay [their eggs] all at once, but little by little, nor do the males release their milt all at once. And all these [things] happen in accord with reason; for just as, within the kind consisting of birds, in some cases they have eggs without impregnation—but few and only a few times, whereas most [are] from copulation—this same [thing] happens also in the case of fish, though [it happens] less.[33] And in both, in as many kinds of them as have the male, the spontaneous [eggs] come to be infertile unless the male sprays [its milt or semen] over [them]. Now in birds, on account of their eggs coming out perfected, [it is] a necessity that this happen while they are still inside [the mother]. But with fish, on account of their [being] unperfected and obtaining growth outside in all [cases], even though the egg hasn't come into being inside from copulation,[34] nevertheless those that are sprayed upon outside are saved, and the milt in the males is expended there. Wherefore also it comes down in a diminished amount at the same time as the eggs in the females do; for they always follow along and spray upon those that are being laid; and so there are males and females and they all copulate, unless in some kind the female and the male are not distinguished, and without the semen of the male none of these sorts [of fish] comes into being. It also contributes to their being deceived

---

33 Cf. 750b9–11, where Aristotle says that this happens "less conspicuously" in the case of fish.

34 At 756a23–24, Drossaart Lulofs would emend the clause *ei kai mē entos ex okheias genētai to ōion*, the reading of the great preponderance of the manuscripts and which I have translated as "even though the egg hasn't come into being inside from copulation." The reading that he proposes, without any ancient support, is *ei kai mēden entos ex okheias genētai gonimon*, which would be translated as "even though nothing inside from copulation has become fertile."

that the coupling of these sorts of fish is fast, so that it escapes the notice of many even among the fishermen; for none of them watches for any such [thing] for the sake of knowing, but nevertheless cou-

756b pling has been seen. For dolphins copulate in the same manner, coming alongside [one another], as do all the fish whose tail hinders [belly to back copulation], but the release in the case of dolphins takes longer, whereas that of these sorts of fish [is] quick. Wherefore

5 not seeing this, but [seeing] the gulpings down of the milt and of the eggs, even the fishermen offer the simple-minded and constantly repeated speech concerning the impregnation of fish that Herodotus the story-teller does, as to how fish become pregnant from gulping down the milt—not understanding that this is impossible.[35] For the channel that goes in through the mouth goes to the

10 stomach, not to the uterus. And [it is] a necessity that what has gone into the stomach become nutriment (for it is thoroughly concocted), but the uterus is manifestly full of eggs, which came from where?

Chapter 6

It is the same way also with the coming into being of birds. For there
15 are some who say that ravens and the ibis have intercourse through the mouth and that, among the quadrupeds, the weasel brings forth [offspring] through the mouth. For both Anaxagoras and some of the other [thinkers] concerned with nature say these [things], speaking very superficially and without [proper] consideration, being deceived concerning the birds as a result of reasoning, from the
20 copulation of ravens rarely being seen, but their uniting with one another by their beaks, which all of the raven-like birds engage in, [being seen] often; and this [is] clear in the case of domesticated jackdaws. And the kind consisting of pigeons also does this same [thing]; but because of their also manifestly having intercourse, on account of this they haven't gotten this tale [told about them]. But
25 the raven-like kind is not [much] given to sexual activity (for it is among the unprolific [kinds]), although even it has before now been seen copulating. And then not to give a reasoned argument as to how the seed reaches the uterus through the stomach, which is al-

---

35 Herodotus, *Histories*, Book Two, Chapter 93.

ways concocting that which comes to be in [it], as [it does] nutriment, [is] strange. These birds too have a uterus, and eggs appear   30
near the diaphragm. And the weasel, just like the other quadrupeds,
has a uterus in the same way as they do—from which how will the
fetus (*embruon*) proceed to the mouth? But because of the weasel
bringing forth very small [offspring], like the other cloven-hoofed
[animals], about which we will speak later, and from its often car-   757a
rying its young with its mouth, [that is what] has brought about this
opinion. Simple-minded and very much deceived [are] also those
who speak about the trochus and the hyena. For many assert about
the hyena, and Herodorus of Heraclea about the trochus, that they   5
have two sets of genitals, [those] of [the] male and [those] of [the]
female, and that the trochus itself copulates with itself, and that the
hyena mounts and is mounted in alternate years. But the hyena has
been seen having one set of genitals; for in some places there isn't
a shortage of [opportunities for] observation. However hyenas have
a line under their tail similar to the genitals of the female. Now both   10
the males and the females have this sort of marking, but males are
captured more [often]; wherefore it has brought about this opinion
in those who observe casually. But concerning such things what has
been said [is] enough.

Chapter 7

But concerning the coming into being of fish, someone might be
perplexed, [wondering] on account of whatever cause neither are   15
the females of the selachian-like [fish] seen spurting out their embryos[36] nor the males their milt, whereas of those that don't bring
forth animals, both the females [are seen spurting out] their eggs
and the males their milt. Responsible [for this is] that the kind consisting of the selachian-like [fish] doesn't in general have much seed.
And further, the females have their uterus by the diaphragm. For   20
the males differ from the males and the females from the females in
a like manner; for the selachian-like [fish] yield little [residue] for
semen. And as for the male kind among those that lay eggs, just as

---

36 The "embryos" mentioned here are presumably their (fertile) eggs. Cf. 731a5–
    6 and 724b16–18.

the females lay their eggs on account of their great number, so those
25   spurt out [their milt], for they have more milt than [is] sufficient
for impregnation; for nature prefers to expend the milt on helping
to make the eggs grow, whenever the female has laid [them], rather
than on their constitution from the beginning. For as has been said
in the speeches above and in the recent ones,[37] the eggs of birds are
perfected inside, whereas those of fish [are perfected] outside. For
30   in a certain manner [fish] resemble those that bring forth larvae, for
those of the animals that bring forth larvae discharge the embryo
[in a] still more unperfected [state]. And for both, both the eggs of
birds and those of fish, the male brings about the perfecting, but
for those of the birds [he does it] inside (for they are perfected inside), whereas for those of the fish [he does it] outside, because of
35   their being discharged outside [still] unperfected, although the
757b   same [thing] results in both cases. And so wind eggs of birds become fertile [through copulation], and those previously made fertile through copulation by a different kind among the males change
their nature into [that of] the one copulating later. And as for
[eggs] of his own [kind] that are not growing if he leaves off his
5   copulation, whenever [the female] is mounted again,[38] it makes
them grow quickly, not however at every time, but if the copulation
comes about before they have changed to the separation of the
white [from the yellow]. But for the [eggs] of fish, nothing of that
sort has been marked off, but the males quickly spray [their milt]
over [them] for their preservation. Responsible [for this is] that
10   these [eggs] are not two-colored; wherefore no such fixed time has
been marked off for these, as [there is] in the case of birds. And
this has come about reasonably; for when the white and the yellow
have been separated from one another, they already have the ruling
beginning from the male (for the male contributes to this [ruling
15   beginning]).[39] So then wind eggs obtain coming into being as far

---

37   718b5–27; 733a3–b16; 751a24–29; 755a11–b1.

38   Although Drossaart Lulofs suspects that there is something wrong with the
text of the beginning of this sentence, it seems reasonably clear, and neither
the manuscript variants that have been noted nor the emendations that have
been proposed seem to me to affect the meaning in a significant way.

39   At 757b13 I read *eis tautēn gar sumballetai to arren* with the great preponderance of the manuscripts, including Z, rather than *tautēn gar sumballetai to*

as it is possible for them; for [it is] impossible for them to be perfected into an animal (for there has to be sense perception), but females and males and all living [beings] have the nutritive power of
the soul, as has been said often.[40] Wherefore the egg itself, considered as an embryo of a plant, is perfect, but considered as [that] of
an animal, [it is] imperfect. So then if there were no male within   20
their kind [namely, that of the birds], they would come into being
as [they do] in the case of the fish[41]—if indeed there is some kind
of such a sort as to generate without a male; but it has been said
about them earlier also that it hasn't yet been observed sufficiently[42]—but as it is, in all the [kinds of] birds, one is male and
the other female, so that insofar as it is a plant [the wind egg] is
perfected (wherefore it doesn't change again after impregnation),   25
but insofar as it is not a plant it is not perfected, nor does anything
different come out of it; for it has come into being neither as [the
egg of] a plant simply nor as [that of] an animal[43] as a result of

---

arren, which Drossaart Lulofs accepts even though it appears in only one
manuscript. With Drossaart Lulofs' reading, instead of "(for the male contributes to this [ruling beginning])," this clause would be translated as "(for
the male contributes this [ruling beginning])." The reading that I have chosen
is admittedly difficult (cf. 737a27–33), but it is consistent with the suggestion
here that what is called "the ruling beginning from the male" is not present
in the eggs from the moment of initial fertilization. Moreover, it seems to me
that since the antecedent of the word "this" is clearly "the ruling beginning
from the male (emphasis mine)," the clause would be redundant on Drossaart
Lulofs' reading.

40 In other words, the ruling beginning in question in this passage is that of the
    sense-perceptive soul. Cf. 741a6–32.

41 At 757b21, Drossaart Lulofs would emend the phrase *epi tōn ikhthuōn*, which
    I have translated as "in the case of the fish," to *epi tinōn ikhthuōn*, though
    without any ancient authority. With Drossaart Lulofs' reading, this phrase
    would be translated as "in the case of some fish." But in either case, the ref-
    erence is presumably to the erythrinus, first mentioned at 741a32–38, and to
    the channa mentioned along with it at 755b20–21.

42 Cf. 741a32–38.

43 At 757b26–27, I read *oute gar hōs phutou haplōs outh' hōs zōiou*, in keeping with
    all the manuscripts, rather than Drossaart Lulofs' *oute gar hōs phuton haplōs
    outh' hōs zōion*, which is supported only by his conjecture regarding the Greek
    source for the Arabic translation. On Drossaart Lulofs' reading, instead of
    "neither as [the egg of] a plant simply nor as [that of] an animal," this phrase
    would be translated as "neither as a plant simply nor as an animal."

coupling. And the eggs that have come into being as the result of coupling and that have been separated into the white [and yellow] come into being in conformity with the one that copulated first;
30   for they already have both ruling beginnings.

Chapter 8

In the same manner the cephalopods, for example cuttlefish and the like, produce their offspring, as well as the soft-shelled [animals], for example crayfish[44] and those of like kind to them; for these too give birth as a result of copulation, and the male has often been seen
35   coupling with the female. Wherefore on this account too those who
758a   assert that fish are all female and that their giving birth is not a result of copulation are manifestly not speaking as inquirers; for to suppose that these [give birth] as a result of copulation but those don't [is] astounding, and if this had escaped their notice, [it would be] a sign of ignorance. The coupling of these [animals], all [of them],
5   takes quite a long time, just as [that of] insects, [and] reasonably so, for they are bloodless, on which account [they are] cold in their nature. Now in cuttlefish and squid the eggs appear [to be] two, on account of the uterus being differentiated and appearing split, whereas that of octopuses [appears to be] one egg. Responsible [for this is] the shape [of its uterus], which [is] round in form and spherical;
10   for its cleavage isn't clear when it is filled [with eggs]. The uterus of crayfish is also split. And all these bring forth an embryo [that is] unperfected on account of the same cause. Now the crayfish-like [animals], the females, deposit their offspring upon themselves (wherefore the females among them have larger flaps than the males,
15   for the sake of protection of the eggs), but the cephalopods [deposit theirs] outside. And the male of the cephalopods sprays [his milt] over the females just as the male fish [spray theirs over] the eggs, and there comes to be a continuous and sticky [mass]. But in the crayfish-like [animals], neither has this sort of thing been seen nor [is it] reasonable; for the embryo[45] is under the female and hard-
20   skinned; and both these and those of the cephalopods obtain growth

---

44 Crayfish (and cuttlefish) are not true fish.

45 Cf. n. 36, above.

outside, just as also those of fish [do]. The cuttlefish that is coming into being is attached to the eggs at its front, for it is possible only there; for it alone has its rear part and its front [part facing] in the same direction. But one should consider the shape of their position, the manner [of shape] that [these parts] have as they are coming into being, from the *Histories*.[46] And so it has been spoken concern-   25 ing the coming into being of the other animals, those that walk and those that swim and those that fly.

Chapter 9

And concerning insects and the shell-skinned [animals] one must speak in accord with the path of inquiry that has been indicated. Let us speak first then about insects. Now it has been said that some   30 of such [animals] come into being as a result of copulation but oth- ers spontaneously,[47] and in addition to these [things, it has been said] that they produce larvae[48] and on account of what cause they pro- duce larvae.[49] For it looks as if nearly all [animals] in a certain man- ner produce larvae at first; for the most unperfected embryo is of this sort, and in all [animals], even those that bring forth animals and those that lay a perfected egg, the embryo obtains its growth at   35 first while it is [still] undifferentiated; and of this sort is the nature of the larva. After this, some [animals] bring forth their embryo [as] a perfected egg, while others [do so as an] unperfected [one], and it becomes perfected outside, as has often been said in the case of fish.   758b And as for those that produce animals inside themselves, [the em- bryo] after its constitution from the beginning becomes in a certain manner egg-like; for its liquid is contained by a thin membrane, just as if someone were to remove the shell of eggs; wherefore they   5 call the destructions of embryos that come about at that time ef- fluxes. Those insects that generate, generate larvae, and those that come into being not through copulation but spontaneously come into being out of this sort of formation at first. For one ought to

---

46   *On the History of Animals* 550a16–26.

47   715a25–b7; 721a2–10; cf. 732b10–14.

48   733a24–25.

49   This may be a reference to 733b10–16.

posit even caterpillars, and the [products] of spiders, as a certain
10 form of larva. And yet some even of these and many of the others
might seem to be like eggs on account of the roundness of their
shape; but one ought not to speak on the basis of their shape nor
on that of their softness or hardness (for the embryos of some become hard), but on that of its changing as a whole and the animal
15 not coming into being out of some part. All the larva–like [embryos],
as they develop and gain in size, finally become like an egg; for the
shell hardens around them and they become motionless at this time.
This [is] clear in the larvae of bees and wasps and in caterpillars.
Responsible for this [is] that their nature as it were lays an egg before
20 its season on account of its own imperfection, as if the embryo while
it is still in [the process of] growing were a soft egg. And it occurs
in the same manner also in the case of the other [insects], all those
that not as a result of copulation come to be in wool or some other
such [materials] and those [that come to be in] water. For all [of
25 them], having become motionless after [exhibiting] the nature of
the larva and with the shell having dried out around [them], after
these [things], when this [shell] has been broken apart, emerge just
like an animal out of an egg, perfected at its third birth, of which
[animals] those that fly [are] greater in number than those that
walk.[50] And what would justly be marveled at by many also happens
30 in accord with reason; for caterpillars, though they take nourishment
at first, no longer take it after that, but what are called chrysalises
by some are motionless, and the larvae of wasps and of bees <take
in nourishment when they are newly born and they are seen to have
excrement; but>[51] after that there come into being what are called
pupae, and they have nothing of that sort. For the nature of eggs,

---

50 At 758b27–28 I accept Drossaart Lulofs' suggested reading, *hōn ta pterōta
pleiō tōn pezōn estin*, which was apparently the text used as the basis for the
Arabic translation, rather than *hōn ta pleista pterōta tōn pezōn estin* in the
manuscripts. I can't give a plausible translation of the reading in the manuscripts, whose literal translation might perhaps be "of which most of those
that fly are [in the class of] those that walk."

51 At 758b32, Drossaart Lulofs suggests that there is a lacuna in the manuscripts
after the words translated as "and the larvae of wasps and of bees." I agree with
his judgment, and the words that I have added in angular brackets are the English translation of his suggestion for filling it, a suggestion that merely restates
what Aristotle wrote about these larvae in *On the History of Animals* 551a29–b2.

when it has reached [its] end, is also without growth, whereas at first   35
it grows and takes in nutriment until it has been differentiated and
has become a perfected egg. Of larvae, some have inside themselves
the sort of [nutriment] from which that sort of residue [i.e., excre-
ment] also comes into being as they are being nourished, <for ex-   759a
ample,>[52] the [larvae] of bees and of wasps, whereas others obtain
it from outside, as do caterpillars and some of the other larvae. And
so why such [insects] come into being in three births, and [the] cause
on account of which they become motionless again, has been said.
Some of them come into being as a result of copulation, just as birds   5
and the [kinds] that bring forth animals and most of the fish,
whereas others [come into being] spontaneously, just as some of the
[plants] that spring up [from the earth].

Chapter 10

The coming into being of bees involves much perplexity. For seeing
that[53] in the case of fish, some [of them], there is a coming into being
of some such sort as generating without copulation, it looks, from   10
what (pl.) appears, as if this happens also in the case of the bees. For
[it is] a necessity that either they fetch their brood from elsewhere,
as some assert—it either springing forth spontaneously or with
some other animal bringing [it] forth—or that they generate [it]
themselves, or that they fetch one [brood] and generate another (for
some say this also, that they fetch only the brood [consisting] of   15
drones); and that they generate either by copulating or without copulation; and that if they copulate, either each kind generates in con-

---

52 At 759a1, I accept Drossaart Lulofs' addition of the bracketed word that I
    translate here as "for example," even though its only ancient support is from
    the Arabic translation.

53 At 759a8, I read *epei*, with the great preponderance of the manuscripts, rather
    than *eiper* with Drossaart Lulofs, who follows what appears, he says, to have
    been the original reading of Z, though a correction by a later hand also reads
    epei. According to Drossaart Lulofs' text, the translation would therefore read
    "if in fact" instead of "seeing that." Drossaart Lulofs' judgement is presumably
    influenced by passages such as 741a32–38 and 757b19–23. However, the read-
    ing that I have chosen finds support from 759b28 below, where Aristotle states
    as a manifest fact that some fish generate without copulation. Cf. 756a22–25.

formity with itself, or some one of them [generates] the others, or one kind [generates] by coupling with another, I mean, for example, that bees[54] come into being from bees coupling and drones from

20 drones and the kings from the kings, or that all the others [come into being] from one, for example, from those that are called the kings[55] and leaders, or [that they come into being] from the drones and the bees; for some assert that the former are males and the latter females, while others [assert] that the bees [are] males and the drones

25 females. But all these [things] are impossible when [we] reason partly from what (pl.) happens in the case of the bees in particular and partly from what (pl.) is more in common with the other animals. For if, not bringing forth [their brood], they fetch [it] from elsewhere, bees ought to come into being even if the bees don't fetch them, in the places from which they fetch the seed. For why, when

30 there will be [bees] if it has been carried away, will there not be [bees] there? For it is no less fitting [that there be bees there], whether [the brood] springs forth spontaneously or from some animal bringing [it] forth. And if the seed were from some different animal, that [one] ought to come into being out of it, but not bees. And further, [it is] reasonable that they fetch honey (for it is food), but absurd

35 that they [fetch] their brood, given that it belongs to others and isn't food. For why would they? For all [animals] that busy themselves

759b over their offspring take pains over what appears [to be] their own brood. But yet neither [is it] reasonable that the bees are females and the drones males; for nature doesn't give a weapon for fighting to any of the females, and whereas the drones are stingless, all the

5 bees have a sting. But neither [is] the opposite reasonable, that the bees are males and the drones females; for none of the males is accustomed to take pains over its offspring, but as it is, the bees do this. And in general since the brood consisting of the drones manifestly comes into being even when no drone is [present], whereas

10 that of the bees [manifestly] doesn't come into being without the kings (wherefore also some assert that only the [brood] of the

---

54 Aristotle uses the same word, *melittai*, both for the kind as a whole and, as here, for those bees we now call the worker bees.

55 Greek uses the masculine word "kings" for what are called in English the queen bees, but it will become clear from Aristotle's account that he regards them as being like females in important respects.

drones is fetched), [it is] clear that they don't come into being as a result of copulation, neither from each of the two kinds itself coupling with itself nor from bees and drones. And that they fetch only this [brood, that of drones, is] impossible on account of what (pl.) has been said as well as [its] not [being] reasonable that there not be a similar sort of state of affairs with regard to all of the kind. But yet neither is it possible that the bees themselves be, some of them, males and the others, females; for in all the kinds [of animals], the female and the male differ. Also they themselves would generate themselves; but as it is, it is manifest that their brood doesn't come into being unless "the leaders are within," as they say. And [an argument] in common both against their coming into being from one another and against their [coming into being] from [copulation with] the drones, [i.e.,] both apart from and with one another, [is] that none of them has ever been seen copulating. But if there were among them one [that is] female and the other male, this would happen frequently. There remains, if they come into being as a result of copulation, that the kings generate [them] by coupling. But even when leaders aren't within, there manifestly come into being drones, whose brood it isn't possible for the bees either to fetch or to generate by themselves copulating. It remains, then, as manifestly happens in the case of some fish, that the bees generate the drones without copulation, being females with respect to generating but having in themselves, like the plants, both the female and the male. Wherefore also they have the organ for fighting; for one ought not to use the name female where there is not a separate male. And if this manifestly happens in the case of the drones, namely, that they come into being not as a result of copulation, [it is] at once necessary for there to be the same account regarding both the bees and the kings, and that they not be generated as a result of copulation. If, then, the brood of the bees manifestly came into being without the kings [being present], it would be necessary for the bees too to come into being out of themselves without copulation. But as it is, since those who are concerned with the care of these animals deny this, it remains that the kings generate both themselves and the bees. And it being the case that the kind consisting of bees is extraordinary and distinctive, so also their coming into being appears to be distinctive. For that the bees generate without copulation would be [something] that happens also in the case

15

20

25

30

35

760a

5

of other animals, but that they not generate the same kind [is] distinctive; for the erythrini generate erythrini and the channae channae. Responsible [for this distinctiveness is] that the bees themselves
10   are also not generated like flies and those kinds of animals,[56] but out
of a kind that is different though akin; for they come into being out
of the leaders. Wherefore their coming into being has a sort of pro-
portionality. For the leaders are similar in size to the drones, but in
15   having a stinger, to the bees; so then the bees resemble them in this
respect, but the drones [do] in respect to size; for [it is] a necessity
that there be some overlapping, if the same kind must not always
come into being out of each (and this [is] impossible; for the kind
would be all leaders). The bees, then, are like them in respect to their
power and in bringing forth [young], whereas the drones [are like
20   them] in respect to size. And if [the drones] also had a stinger, they
would be leaders. But now this [part] of the perplexity remains; for
the leaders resemble both [kinds] at the same time, in having a
stinger, the bees, and in size, the drones.[57] And [it is] necessary for
the leaders too to come into being out of some [kind]. So since [it
25   is] neither out of the bees nor out of the drones, [it is] necessary for
them to generate themselves as well. Their cells come into being
last, and [they are] not many in number. So that [what] happens [is]
that the leaders generate themselves and generate also some other
kind (this is that of the bees), and that the bees, while they generate
30   some other [kind], the drones, no longer generate themselves, but
this has been taken away from them. And since what is in accord
with nature always has an order, because of this [it is] necessary for
even generating some other kind to have been taken away from the
drones. Which is also manifestly the case; for they themselves come
into being, but they don't generate anything else, but coming into
35   being has reached its limit in the third number [of the series]. And
760b   in this way, then, it is arranged beautifully by nature so that the kinds

---

56 In other words, they are not generated spontaneously. Cf. 721a2–9; and *On
the History of Animals* 539b7–14.

57 Drossaart Lulofs would delete this sentence and the one before it. But they
appear in all the manuscripts, and I choose to include them. As for why these
resemblances are said to be perplexing, it might have been expected that the
drones would both lack a stinger and also not resemble the leaders in size, and
thus be at the furthest remove from them in these respects, as they will be
said to be at the furthest remove with respect to the power to generate.

may remain always [in] being and not one [of them] fail, even
though they don't all generate. [It is] reasonable also that the following happen, that in good seasons there come to be [much] honey
and many drones, but in rainy seasons on the whole a large brood.
For wet conditions produce more residue in the bodies of the lead-   5
ers, whereas good seasons [produce it] in those of the bees; for being
smaller in size, they are more in need of a good season. And [it is]
well also that the kings, having been made as it were for bearing
young, remain inside, released from the necessary tasks, and that
they be big, with their body as it were having been constituted for   10
bearing young; and that the drones [be] idle, inasmuch as they have
no weapon for fighting it out over food and on account of the slowness of their body. And the bees are in the middle between both in
size[58] (for in this way they are useful for working), and industrious,
as providing food for both children and fathers. And also in agree-   15
ment [with this account] is that they attend upon the kings, because
their coming into being, [namely,] that of the bees, is from these (for
if nothing of this sort were the case, the facts regarding their lead-
ership wouldn't be reasonable), and that they allow the ones to do
no work, as [being] their parents, whereas they punish the drones,
as [being] their children; for it is more beautiful to punish one's chil-   20
dren and those who have no work. And that the leaders, who are
themselves few, generate the bees, [who are] many, looks as if it happens in nearly the same way as the coming into being of lions, who
after having generated five at first, later on generate fewer, and fi-

---

58 At 760b13, I read *hai de melittai mesai to megethos eisin amphoin*, in keeping
    with all the manuscripts, even though the reading contradicts Aristotle's ear-
    lier claims that the drones resemble the leaders in size and that the bees are
    smaller (760a12–15; b6–7). A plausible modern emendation, accepted by some
    editors, is *hai de melittai meious to megethos eisin amphoin*, which would be
    translated as "and the bees are lesser in size than both." Drossaart Lulofs
    would preserve the word *mesai* but delete *to megethos*, leading to a translation
    "and the bees are in the middle between both," without any reference to size.
    I accept the manuscript tradition without much confidence that it is correct,
    but I find it hard to see how it could have arisen if it isn't what Aristotle
    wrote. And given that Aristotle has just put a good face on the puzzling exis-
    tence of drones, as he understood them, by claiming that is well that they be
    idle (cf. 741b4–5; 744a36–37), it doesn't seem out of the question that he
    might have deliberately contradicted what he knew to be the facts regarding
    size in favor of what might seem to be a more rational order. Cf. n. 57, above.

25   nally one, then none. The leaders at first [generate] a multitude, and
     later a few [of the same kind as] themselves, and nature [made] the
     brood of the latter a smaller [one], but since it took away from their
     multitude, it gave them size in return. So then on the basis of reason
     the [things] that have to do with the coming into being of bees ap-
     pear to be in this way, and on the basis of what (pl.) is thought to
30   happen with regard to them; however, what (pl.) does happen has
     certainly not been grasped sufficiently, but if it ever is grasped, then
     one must trust sense perception more than arguments, and argu-
     ments if what (pl.) they show is in agreement with what (pl.)
     appears. And a sign in support of their not coming into being from
     copulation [is] also that the brood appears small in the cells of the
35   comb. But as many of the insects as are generated from copulation,
761a while they couple for a long time, bring forth quickly and [what they
     bring forth is something] larva-like that has size.[59] Concerning the
     coming into being of the animals akin to them, for example, hornets
     and wasps, it is in a certain manner nearly the same for all; but what
5   is extraordinary has been taken away, reasonably so; for they don't
     have anything divine, as [does] the kind consisting of bees. The so-
     called "[mother] wombs" generate and they mold the first of the
     cells, but they generate after being impregnated by one another; for
     their coupling has often been seen. How many differences each of
10   such kinds has, either from one another or from bees, must be stud-
     ied on the basis of what has been recorded in the *Histories*.[60] And it
     has been spoken concerning the coming into being of the insects, all
     [of them], but it must be spoken about the shell-skinned [animals].

Chapter 11

     What (pl.) is involved in the coming into being of these is partly
15   like and partly unlike [what is involved in that of] the others. And
     this happens reasonably; for compared with the animals they are
     like plants, and compared with plants, [like] animals, so that in a
     certain manner they manifestly come into being from a seed, but

---

59 Drossaart Lulofs brackets this sentence and the one before it, since he suspects
    that they don't belong here.

60 *On the History of Animals* 627b23–629a29.

in another manner not from a seed, and in a way spontaneously but in a way from themselves, or some of them in the one manner but others in the other. And because of their nature being set over against [that of] the plants, because of this, none of the shell-skinned [animals] comes into being in the earth, or [only] some small kind, for example, that of the snails and if [there is] any other, albeit a rare one, of that sort, but in the sea and the wet [places] like it, many [of them come into being] and with every sort of shape. But the kind consisting of plants [is] small and altogether nothing, so to speak, in the sea and the [places] of that sort, whereas in the earth all such [beings] come into being; for the nature [that these two kinds] have [is] analogous, and to the extent that the wet [is] more conducive to life than the dry, and water than earth, to that extent the nature of the shell-skinned [animals] stands apart from that of the plants, since as the plants [are] in relation to the earth, so the shell-skinned wish to be, at any rate, in relation to the wet, with plants being, as it were, land shellfish and shellfish, as it were, aquatic plants. And also on account of such a cause the [animals] in the wet [place] are more varied in shape than those in the earth; for the wet has a nature that is easier to mold, and not much less bodily, than that of earth, and of this sort [are] especially the [animals] in the sea; for fresh [water], though sweet and nutritious, [is] less body-like and [is] cold. Wherefore, the [animals] that are bloodless and not hot in their nature, for example, the shell-skinned and the cephalopods and the soft-shelled (for all these are bloodless and cold in their nature), do not come into being in lakes or in the fresher of the salty [waters], except to a relatively small extent, but they come into being in lagoons and near the mouths of rivers; for they seek warmth and nutriment at the same time, and the sea is wet as well as [being] much more body-like and hot in its nature than fresh [water], and it has a share of all the parts, wet and *pneuma* and earth, so as also to be a share in all the animals that come to be in accord with each [of these parts] in the places [proper to each of them]. For someone might set plants down [as] belonging to earth, and the aquatic [animals] to water, and the land [animals] to air; but the more and less and nearer and farther make a great and wondrous difference. But one ought not to seek the fourth kind in these places; and yet there wishes, at any rate, to be something in accord with the ordered position of fire; for this is counted [as] the fourth of the

20

25

30

35
761b

5

10

15

135

bodies. But fire always is manifest [as] having a form that is not its own, but [as being] in a different one among the bodies; for what

20   has been set on fire [is] manifestly either air or smoke or earth. But one has to seek the kind that is of that sort on the moon; for this manifestly has a share of [what is at] the fourth remove. But concerning these [things] there would be another account. The nature of some of the shell-skinned [kinds] is constituted spontaneously,

25   whereas [that] of some [is] by their emitting a certain power from themselves, though these too often come into being from a spontaneous formation. Now one needs to grasp the comings into being of plants. For some of these come into being from a seed, some from slips that are planted out, and some, for example, the kind consist-

30   ing of onions, by coming forth as offshoots. Now mussels come into being in this manner; for smaller [ones] are always growing by the side of the ruling beginning. Whelks and purple-fish and those that are said to make "honeycombs" [61] emit slimy fluids, as from a spermatic nature. But one must hold [that] none of these [is] seed, but that in the manner that has been stated they share in the likeness to

35   plants; wherefore also a multitude of such [animals] comes into
762a being whenever there has once come into being something [of the same kind]. For while it is the case that all these also come into being spontaneously, they are constituted more [readily], in accord with reason, when [some] have also been present already. For it is reasonable that some residue come into being as a surplus close by each [part] of the ruling beginning from which each of the [offshoots]

5   that grows beside [it] comes forth. And since the nutriment and the residue from it have nearly the same power, it makes sense that those that make "honeycombs" should have a being that is similar to the formation from the beginning; wherefore [it is] reasonable that [still others of the same sort] should come into being also out of this. As for as many [of the shell-skinned animals] as neither come forth as offshoots nor make "honeycombs," the coming into being of all

10   these is spontaneous. And all [beings] that are constituted in this manner, both in earth and in water, manifestly come into being along with rotting and as rainwater is mixed in; for when the sweet [rainwater] is separated off into the ruling beginning that is being

---

61 Regarding these so-called "honeycombs," see *On the History of Animals* 546b18–547a4.

constituted, what remains as residue gets that [other] sort of aspect. However nothing comes into being by rotting, but by being con- cocted; for rotting, i.e., what is rotten, is a residue of what has been   15 concocted; for nothing comes into being from all [of its material], just as it doesn't in the case of the [things] that are fashioned by art—for there would be no need to make [anything]—but as it is, in the one case art removes what (pl.) is useless, and in the other case nature [does]. Animals and plants come into being in earth and in [what is] wet on account of water being present in earth and *pneuma* in water and soul-ish heat in this, [in] all [of it], so that in a   20 certain manner all [things] are full of soul. Wherefore [animals and plants] are constituted quickly whenever [this or these] have been enclosed. And [this or these] get enclosed and, as the body-like flu- ids are heated, there comes to be as it were a foamy bubble. Now the differences as to what is being constituted being more precious in kind or more worthless depend on the enclosure of the soul-ish   25 ruling beginning. And both the places [in which it happens] and the body that is being enclosed [are] responsible for this [outcome]. In the sea there is much [that is] earthy; wherefore from that sort of constitution the nature of the shell-skinned [animals] comes into being, when what is earthy hardens all around and solidifies with   30 the same solidity as bones and horns (for these are unmeltable by fire), whereas the body that has the life gets enclosed inside. Of such [animals] only the kind consisting of snails has been seen coupling. But whether or not their coming into being is the result of the cou- pling has not yet been seen sufficiently. Someone wishing to inquire   35 correctly might ask what it is in such [beings] that, corresponding   762b to the material ruling beginning, is being formed. For in females this is a certain residue of the animal, by setting which in motion, [this residue] being potentially of such a sort as that from which it came, the ruling beginning from the male perfects the animal. But what should we say [is] of that sort here, and from where, and what,   5 [is] the ruling beginning, corresponding to the male, that sets [it] in motion? Now one must grasp that even in the animals that generate, out of the nutriment that comes in, the heat in the animal, by sepa- rating off and concocting, makes the residue, the ruling beginning of the embryo. Likewise also in the plants, except that in these and in some of the animals there is no additional need of the male's rul-   10 ing beginning (for they have it mixed within themselves), whereas

137

the residue of most animals does have further need [of it]. And in some, water and earth [are] nutriment, whereas in others [it is] what (pl.) comes out of these, so that the very [thing] that the heat in the animals [i.e., those that generate,] produces out of their nutriment,

15 this the seasonal heat in the environment aggregates by concoction and constitutes out of sea and earth. And that [portion] of the soulish ruling beginning that is enclosed or separated off in the *pneuma* produces an embryo and imparts motion. Now the constituting of the plants, those that come into being from spontaneity, is [in all cases] of the same form; for they come into being out of some part

20 [of something], and some [of that part] becomes the ruling beginning, while some [becomes] the first nutriment for the [shoots] that are growing out.[62] But some of the animals[, that is, of those that come into being from spontaneity,] are brought forth as larvae, [some] even of the blooded [ones],[63] for example, a certain kind of

---

62 Cf. 715b25–30; 732a28–32.

63 At 762b21–22, Drossaart Lulofs, in keeping with Z as well as the Arabic translation, adds the words *kai tōn anaimōn hosa mē apo zōiōn gignetai* before the words that I have translated as "[some] even of the blooded [ones]." They are absent, however, from the preponderance of the manuscripts, and I have chosen to omit them. If they were included, instead of "But some of the animals[, that is, of those that come into being from spontaneity,] are brought forth as larvae, [some] even of the blooded [ones]," the beginning of the sentence would be translated as "But some of the animals, both of the bloodless [ones], as many as do not come into being from animals, and of the blooded [ones], are brought forth as larvae." I omit these words primarily because it is clear from the immediate context (*hē men . . . ta de*; 762b18–21) that "the animals" mentioned in this sentence include only those that come into being from spontaneity, i.e., not from other animals, so that to add them would be superfluous. Moreover, these words create the false impression that only those among the bloodless animals that come into being spontaneously are brought forth as larvae (cf. 758b6–7). I suspect that the words were added by some early scribe who was bothered by the apparent suggestion that only "some" of the animals that come into being from spontaneity are brought forth as larvae, given that Aristotle will soon say that, while we see the spontaneous coming into being of animals from larvae, we don't see it from eggs nor does such an emergence of live young even seem possible (762b31–763a1; 763a5–8). Consider, however, that the Greek word *gignomenōn*, which I translated at 762b18–19 (and, though it wasn't repeated explicitly, in the first bracketed addition to this sentence) as "that come into being," is a progressive participle, which though it is usually a present tense, as I have translated it, can also be a past tense, i.e., the imperfect. Compare the progressive infinitive *gignesthai* at 762b30, along with the

mullets and [some kinds] of other river fish, and further, the kind
consisting of eels. For all these, even though the nature they have
is [one] with little blood, are nevertheless blooded, and they have a   25
heart, the blood-like ruling beginning of the parts. And the so-called
entrails of the earth, in which the body of the eels comes into being,
have the nature of a larva. Wherefore also someone might suppose
with regard to the coming into being of human beings and quadrupeds, if indeed they ever used to come into being [as] earthborn,
as some assert, that they came into being in one of two ways, either   30
as a larva that was constituted at first or out of eggs.—For [it is]
necessary either that they had the nutriment for growth in themselves (and an embryo of that sort is a larva) or that they got it from
elsewhere, and this either from the [female] that generated [them]
or from a part of the embryo; so that if the one [is] impossible,   35
[namely,] for it to flow out of the earth as among the other animals   **763a**
[it flows] out of the mother, [it is] necessary [if they got it from else-
where] that they got it from a part of the embryo; and we say that
coming into being of that sort is out of an egg.—So then if indeed
there was some beginning of coming into being for all the animals,
that [it is] reasonable that it was one or the other of these two [is]
manifest; but there is less reason [in the view that it was] out of eggs;   5
for of no animal do we see a [spontaneous] coming into being of
that sort, but [we do see] the other sort, both of the blooded [ones]
that have been mentioned and of the bloodless [ones]. And of such
a sort are some of the insects as well as the shell-skinned [animals],
with which [our] account is concerned; for they do not come into
being out of some part, like those brought forth as eggs, but they
effect their growth in the same manner as larvae, for larvae grow to-   10
ward the upper [part] and the ruling beginning; for the nutriment
for the upper [parts] is in the lower [part]. In this respect, indeed,
they are also similar to those [that come into being] out of eggs, except that those consume [the nutriment] entirely, whereas in the
ones brought forth as larvae, when the upper part has grown out of   15
the substance in the lower part, the lower [part] is then articulated
out of what is left over. And responsible [for the similarity is] that

---

progressive infinitives *ekhein* and *lambanein* in the following sentence, which
are all naturally translated as past tense verbs ("that they came into being";
"that they had"; and "that they got").

in all [animals], later as well, the nutriment comes to be in the part
below the diaphragm. That the larva-like [beings] effect their growth
in this manner [is] clear in the case of the bees and the [others] of
20   that sort; for in the beginning their lower part is big and the upper
[part] small. And in the case of the shell-skinned [animals], the [way
things proceed] with regard to their growth is in the same manner.
This [is] manifest in the case of those that are spiral-shelled from
their convolutions; for as they are growing, there always come to be
more [of these][64] toward the front and what is called the head. And
25   so, then, it has pretty much been stated what the manner of coming
into being is, both of these and of the others that are spontaneous[ly
generated]. And that all the [kinds of] shell-skinned [animals] are
constituted spontaneously [is] manifest from the following sorts of
[things]: that they come into being on the side of boats when the
foamy slime [on them] rots, and [that] in many places, where nothing
of the sort was present before, later, when the place had become
30   muddy because of a lack of water, there have come into being the
shell-skinned [animals] that are called lake-oysters, for example,
when a naval fleet cast anchor at Rhodes and earthenware jars were
thrown out into the sea, after time had passed and mud had collected
around them, oysters kept being found in them. And that such [ani-
mals] don't emit anything generative from themselves, a piece of
763b   evidence [is that] when certain Chians brought [some] of the oysters
alive from Pyrrha in Lesbos and released them in certain places in
the seas [that are] narrow and have [currents] flowing together,[65]

---

64   At 763a23, I read *pleious*, in keeping with the great preponderance of the manuscripts, rather Drossaart Lulofs' *meizous*, which is supported only by the Latin
translation by William of Moerbeke and perhaps the Arabic translation. According to Drossaart Lulofs' text, instead of "there always come to be more [of
these]," the translation would read "[these] always come to be bigger."

65   At 763b3 I accept Drossaart Lulofs' decision to read *homorrous*, even though
it is a modern emendation. This emendation receives some small support
from Z, which apparently reads *homorous* (meaning "sharing a boundary,"
which makes no sense in the context). Most of the manuscripts read *homoious*,
which can mean "similar," and which was apparently interpreted by some ancient translators to mean "muddy," i.e., "similar to the places where these oysters were seen to come into being spontaneously." But it seems more likely
that the passage refers to places that are dissimilar to those, i.e., to places where
the ocean currents would prevent the formation of mud and hence the spontaneous coming into being of these oysters.

they became no more numerous over time, but grew greatly in size. And their so-called eggs contribute nothing toward coming into being, but are a sign of good feeding, like fat in the blooded [animals]; wherefore they also become flavorful for eating around these times. A sign [of this is] that those [of the shell-skinned animals] such as pinnae and whelks and purple-fish always have [them], except that [they are] sometimes bigger and sometimes smaller. Others, for example, scallions and mussels and what are called lake-oysters, don't [have them] always, but they have them in the spring, whereas they wither as the season advances and finally disappear entirely; for the [spring] season itself is advantageous for their bodies. And in others, for example, sea-squirts, there occurs nothing [that is] clearly seen of this sort. But as for the particulars concerning these and the places in which they come into being, let it be studied from the *History*.[66]

5

10

15

---

66 *On the History of Animals* 527b36–531a8; 546b15–548a22; 590a18–590b9; 599a10–20.

Book Four

Chapter 1

Now then it has been spoken concerning the coming into being of   20
the animals, both in common and separately concerning all [of
them]. But since in the most perfect among them the female and
the male are separated, and [since] we assert that these powers are
[the] ruling beginnings of all, both of animals and plants, but that
some have them unseparated whereas others [have them] separated,   25
one must speak about the coming into being of these first; for the
female and the male are distinguished while they are still imperfect
in their kind. But it is disputed whether even before the difference
is clear to our sense perception, one is female and the other male,
acquiring the difference within the mother or even earlier. For some,   30
for example, Anaxagoras and others of those who speak about nature, say that this opposition is in the seeds straightaway; for [they
assert that] the seed comes into being from the male, whereas the
female provides the place, and that the male is from the right [tes-
ticle] and the female from the left, and that the males are on the
right side of the uterus and the females on the left. Others, like   764a
Empedocles, [assert that the opposition comes about] in the womb;
for he asserts that the [seeds] that come into the uterus [when it is]
hot become males and those [that come] into [it when it is] cold, females, and that the flow of the menses is responsible for the heat   5
and the coldness, [it] being colder or hotter, i.e., of older date or
more recent. Democritus of Abdera asserts that the differentiation
of the female and the male comes into being in the mother, not,
however, that one becomes female and the other male on account
of heat or coldness, but [that the outcome depends on] whichever   10
[parent's] seed, coming from the part by which the female and the
male differ from one another, gains the mastery. For in truth Empedocles conceived of this too casually, supposing that they differ from

one another only in coldness and heat, though he saw that the parts as a whole have a great difference, [namely,] the [difference] between

15 the [male] genitals and the uterus. For if, supposing that the animals had been molded, one having the parts of the female and the other those of the male, they were put into the uterus as into an oven, the one having a uterus [put] into a hot one and the one not having [a uterus put] into a cold one, the one not having a uterus will [on this

20 view] be female and the one having [one], male. But this [is] impossible. So that in this respect, at any rate, what Democritus says would be better; for he seeks the difference [between the two cases] of this coming into being and tries to state [it]—whether finely or not finely [is] another argument. Then again also, if heat and coldness [is] responsible for the difference of the parts, those who speak

25 in that [other] way ought to have said this; for this is, one might say, to speak about the coming into being of male and female; for they differ manifestly in this respect. And [it is] no small task to obtain the cause concerning the coming into being of these parts from that beginning, as if [it were] necessary for it to follow from the animal

30 being cooled that this part, which they call a uterus, come into being, and from it being heated that it not come into being. And [it is] the same way also concerning the parts that serve for intercourse; for these too differ, as has been said earlier. And further, twins, a female and a male, often come into being together in the same part

35 of the uterus, and this we have observed sufficiently from dissections in all those that bring forth animals, both in the land animals and in the fish; concerning which [sorts of twins], if [Empedocles] didn't see [any], his mistake was reasonable in stating this cause, but

764b if he did see [some], [it is] absurd for him still to have held that heat or coldness of the uterus [was the] cause. For [on that view] they would both come into being either female or male, but as it is, this [is] not [what] we see happening. And when he says that the parts of the [animal that is] coming into being are separated asunder (for

5 he asserts that some are in the male [parent] and others in the female, wherefore also[, he asserts,] they desire intercourse with one another), [it is] necessary [for him to say] also that the [complete] magnitude of these sorts of [parts] is divided and there is a coming together, but not on account of cooling or heating. But concerning such a cause of the seed there would probably be many [things] to

10 say; for in general the manner of cause [that he asserts] looks fab-

ricated. But if concerning seed it is as we have in fact stated, i.e.,
[if] it doesn't come from all [the body] and [if] what [comes] from
the male provides no material at all for the [animals] that are coming
into being, one must oppose in the same manner both this one and
Democritus and anyone else who may happen to speak in this way.   15
For neither is it possible for the body of the seed to be, [being]
separated asunder, with it partly in the female [parent] and partly
in the male, as Empedocles asserts when he says,

> But separated asunder is the nature of the limbs, one part in
> the man's...[1]

nor [is it possible] for it to be secreted in its entirety from each of
the two and for this to become female and that male because of a   20
particular part mastering another part. But in general [to say] that
the superiority of the part [namely, of the uterus], by gaining mastery, makes a female [is] better than, not having given it any thought,
to make heat alone responsible; however, the fact that at the same
time the form of the genitals also turns out different [from the
male's] needs an argument as to why these [parts] should always accompany one another. For if because they are near, each of the re-   25
maining parts would also have to accompany [them]—for one of
those that gain victory is near another—so that [the offspring] would
at the same time be female and resembling the mother or male and
[resembling] the father. Further, [it is] absurd also to suppose that
only these parts have to come into being, and not that the body as a
whole [has to] have changed, above all and in the first place the blood
vessels, around which, as around a framework, is placed the body   30
consisting of flesh. [It is] not reasonable that these come into being,
being of a particular sort, on account of the uterus, but rather that
the uterus [do so] on account of those; for while each of the two is a
receptacle for a particular [sort of] blood, that consisting of the blood
vessels is [formed] earlier. And [it is] necessary always for the ruling
beginning that sets in motion to be earlier and [to be] responsible,   35
by being of a particular sort, for the coming into being. So then the
difference from one another of these parts in females and males [is
something that] results, but one ought not to suppose that this [dif-

---

1   Empedocles, fr. 63 (Diels/Kranz); cf. 722b12.

765a ference] is a ruling beginning or a cause, but rather [that] a different [one is], even if no seed is secreted either from the female or from the male, but the seed that comes into being[2] is constituted in any way whatever. And the same argument as the one against Empedo-

5 cles and Democritus [serves] also against those who say that the male is from the right [testicle] and the female from the left. For if the male contributes no material, those who speak in that way would be saying nothing; and if, as they assert, it too contributes [material], [it is] necessary to oppose [them] in the same way as [we opposed]

10 the account of Empedocles that distinguishes the female from the male by heat and coldness of the uterus. They do this same thing, distinguishing by the right and the left, even though they see the female and the male differing also by entire parts, among which, [to take] the body that consists of the uterus—on account of what cause will it belong to those from the left but not belong to those from the

15 right? For if it comes [from the left] but doesn't have this part, there will be a female not having a uterus, and [there might also be] a male having [one], if it so happens. And further, as was said earlier as well, a female has been seen in the right part of the uterus and a male in the left and both in the same part, and this not only once but many

20 times, or the male on the right and the female on the left; but no less do both come into being on the right.[3] And some are persuaded in nearly the same manner as these [are] and say that when [animals] copulate with the right testicle or the left one tied off, it results that

25 they produce males or produce females;[4] indeed, Leophanes used

---

2 At 765a3, Drossaart Lulofs would delete *to sperma* ("the seed"), even though it is present in all the manuscripts. If this deletion were accepted, the translation, instead of "the seed that comes into being," would be "that which comes into being." Naturally, that would be an easier reading; but it is hard to see why a scribe would have added these words.

3 At 765a20–21, Drossaart Lulofs would delete the words translated as "or the male on the right and the female on the left; but no less do both come into being on the right," even though they are present in all the manuscripts. They were apparently not present, however, in the manuscript used as the basis for the Arabic translation.

4 At 765a24, the Greek text used as the basis for the Arabic translation apparently read (in English translation) "they produce females or produce males," instead of "they produce males or produce females." Drossaart Lulofs considers this to be the superior reading, as do I, but he doesn't incorporate it into his text.

to speak in this way. And some assert that this same [thing] results
in the case of those with one or the other testicle cut off, not saying
[what is] true, but divining what will result on the basis of likelihoods
and forming a preconception that it is in this way before seeing it
come about in this way, and furthermore, not recognizing that these
parts in animals contribute nothing to generation (*genesin*), [namely,]   30
the generating of males and generating of females. A sign of this [is]
that many of the [kinds of] animals are themselves female and male,
and they generate some that are female and others that are male,
though without having testicles, as [is the case with] those that are
footless, for example, the kind consisting of fish and that of snakes.
Now to suppose heat and coldness, and the secretion coming about   35
from the right or from the left, [to be] a cause of the male and the   765b
female does have some reason [to it]. For the right [parts] of the body
are hotter than the left [ones], and the seed that is concocted is hotter,
and of that sort is what is compacted, and what is more compacted
is more fertile. But to speak in this way is to [try to] lay hold of the   5
cause from too far away, whereas [what] is needed [is] to approach,
[starting] from what (pl.) is admitted, as near as one can to the first
causes. Now then, concerning the body as a whole and its parts, what
each is and on account of what cause, it has been spoken earlier in
other [speeches].[5] But since the male and the female have been distinguished by a certain power and by a lack of power—male, namely,   10
[being] that which is able to concoct and to constitute and expel a
seed that contains the ruling beginning of the form (and by ruling
beginning, I don't mean the sort out of which, as material, there
comes into being [an offspring] of the same sort as the generator, but
that which first sets in motion, whether it is able to do this inside itself or inside another), and female [being] that which receives [a   15
seed] but is unable to constitute [one] and expel [it][6]—further, if all
concoction is effected by means of [what is] hot, [it is] necessary for
the males among the animals to be hotter than the females. For [it
is] on account of coldness and lack of power [that] the female has
more blood in some regions [of the body], and the cause on account
of which some suppose that the female is hotter than the male, the   20

---

5   The reference is to *On the Parts of Animals*, Books Two, Three, and Four, and
     perhaps also to Book One of this treatise. Cf. 715a1–18.

6   Cf. 716a2–23; 728a17–25ff.

discharge of the menses, is a sign of the very opposite. For blood is hot, [they argue,] and what has more is more so. And they take it that this occurrence comes about on account of an excess of blood and heat, as if it is admissible that everything is equally blood if only it
25   is liquid and blood-like in color, and [as if] there doesn't come to be less, and purer, [blood] in the well nourished. But they suppose that, just as with the residue in the bowels, more rather than less is a sign of a hotter nature. And yet it is the opposite; for just as in the workings that have to do with crops, out of much [that is] the first nutri-
30   ment, little, the useful [part], is separated off, and in the end the last [product] is no portion at all compared to the first amount, so again also in the body, out of all the nutriment, after the parts receive [it] in turn for their workings [on it], an extremely small [product] comes into being. And in some this is blood, and in others what is
35   analogous. And since one [kind] has the power and the other lacks the power to expel the residue [in a] pure [form], and [since] there
766a is some organ for every power, both for the one that brings the same [thing] to completion in a worse manner and for the one [that does so] in a better one, and [since] the female and the male—given that "that which has power" and "that which lacks power" are meant in many ways—are opposed in this manner, [it is] thus a necessity that there be an organ both for the female and for the male; now for the
5   one it is the uterus and for the other its genital area. And nature gives the power to each at the same time as the organ; for [it is] better in this way. Wherefore each of the regions comes into being at the same time as its secretions and its powers, just as also sight is not perfected without eyes nor an eye without sight, and bowels and bladder [are perfected] at the same time as it becomes possible for their residues
10   to come into being. And since that out of which [something] comes into being is the same as [that out of which] it grows—and this is its nutriment—each of the parts would come into being out of the sort of material and the sort of residue that it is fit to receive. But furthermore, to the contrary, it comes into being, as we assert, out of its opposite in a sense. And third, in addition to these [things], one must take it that if indeed destruction [is] into the opposite, [it is]
15   also a necessity for that which is not mastered by what is doing the fashioning to change into its opposite. And with these [things] being assumed, perhaps [the] cause on account of which the one [is] female and the other male might now be more manifest. For whenever the

150

ruling beginning doesn't gain the mastery and doesn't have the power to concoct on account of a lack of heat and doesn't bring [the material] into the distinctive form that belongs to it but is defeated in this respect, [it is] a necessity [for that] to change into its opposite. And opposite to the male [is] the female, and in this respect, in which the one is male and the other female. And since [the female] has a difference in power, the organ it has [is] also different, so that it changes into what is of that sort. And when one vital part changes, the animal's constitution as a whole differs greatly in form. It is poss-ible to see [this] in the case of eunuchs, who, with one part having been mutilated, diverge so greatly [as they do] from their original shape and fall little short of the female in form. And responsible for this [is] that some of that parts are ruling beginnings, and when a ruling beginning is moved [i.e., changed], [it is] a necessity that many of the [things] that follow from [it] undergo change. If, then, the male is a certain ruling beginning and responsible [i.e., a cause]— and [a being] is a male insofar as it has a certain power, and a female insofar as it lacks power—and the distinguishing mark of the power and the lack of power is being able to concoct or not being able to concoct the last [stage of] nutriment, which in the blooded [animals] is called blood and in the others what is analogous [to it], and that which is responsible for this [is] in the ruling beginning and the part containing the ruling beginning [that consists] of the natural heat, [it is] therefore necessary, in the blooded [animals], that a heart be constituted and that the [offspring] which is coming into being will be either male or female, and, in the other kinds that have the female and the male, that which is analogous to the heart. And so the ruling beginning of the female and of the male, and their cause, is this and in this [part]. But it is only at that point female and male when it also has the parts by which the female and the male differ; for [it is] not by virtue of any part whatsoever [that] it is male or female, just as [it is not] a seeing [being] and a hearing [being]. Taking up [our account] again, we say that it has been assumed that the seed is the last residue of nutriment (and by "last" I mean that which is carried to each [part], wherefore also that which is generated resembles that which generated [it]; for it doesn't make any difference whether [we say that] it comes from each of the parts or [that it] goes to each, though the latter is more correct). And the seed of the male differs in that it has within itself a ruling beginning of such a sort as to ini-

20

25

30

35

766b

5

10

tiate motion, in the [developing] animal as well, and to thoroughly concoct the last nutriment, whereas the [seed] of the female [has]
15 only material. And so when [the male seed] gains the mastery, it draws [the material] to itself, but if it is mastered, [the material] changes into the opposite or [comes] to destruction. And opposite to the male [is] the female, and [it is] female because of a lack of concoction and the coldness of its blood-like nutriment. And nature gives to each [sex] the part that is receptive of its residues. And seed
20 [is] a residue and this, in the hotter [animals], i.e., males, among those that are blooded, is moderate in amount, wherefore the parts receptive of this residue are channels in the males. But in the females, on account of a lack of concoction, [it is] a large amount, [and] bloody, (for [it is] not thoroughly worked up), so that [it is] necessary that there also be a certain part receptive [of it], and that this be unlike [the channels in males] and have [greater] size. Wherefore the
25 nature of the uterus is of such a sort. And by this part the female differs from the male. So then on account of what cause the one comes into being female and the other male has been said.

Chapter 2

The facts [are] evidence for what (pl.) has been said. For the young [are] more prone to produce females than are those in their prime,
30 and those that are older [are also] more so; for in the former the heat [is] not yet perfected, and in the latter it is failing. And of bodies, those that are wetter and more womanish are more prone to generate females, and seeds that are wet [more so] than those that are coagulated. For all these [things] come to be on account of a lack of natural heat, and [so does] the producing of males when there are north winds more than when there are south winds. For bodies are wetter
35 when there are south winds,[7] so that [they are] also more abundant in residue. And the more residue, [the] harder [it is] to concoct;

---

7   At 766b34b, Drossaart Lulofs includes the words *hugrotera gar ta sōmata notiois*, translated here as "For bodies are wetter when there are south winds," even though they are not present in the manuscripts. Words to this effect were apparently present in the manuscript used as the basis for the Arabic translation, and something of the sort seems necessary to make the sentence grammatical.

wherefore the seed in the males is wetter [then], as is the discharge   767a
of the menses in the women. Also, that the menses, in accord with
nature, come to be more when the moon is waning results from the
same cause. For this time of the month [is] colder and wetter because
of the waning and the failing of the moon; for whereas the sun makes   5
winter and summer in the year as a whole, the moon [does so] in the
month (and this not because of its turnings,[8] but the one happens
when its light is increasing and the other when it is waning). And
also, the shepherds assert that it makes a difference in terms of generating females or generating males not only whether the copulation
happens to take place when there are north winds or south winds,   10
but even if, while copulating, they look to the south or to the north.
Such a small turn of the scale[, they say,] sometimes becomes a cause
of cold and heat, and these of coming into being. In general, then,
the female and the male are distinguished with regard to generating
males and generating females on account of the causes that have been   15
stated; however, there is also need of due measure in relation to one
another; for all the [beings] that come into being in accord with art
or nature are [what they are] by virtue of a certain ratio. And what is
hot, if it gains too much mastery, dries out liquids, but if it greatly
falls short, it doesn't constitute [them], but it needs to have this ratio,
that of the mean in relation to what is being fashioned; and if it   20
doesn't, just as in [meats] that are being boiled, too much fire burns
[them] and too little doesn't boil [them], and in both ways it turns
out that what is coming into being isn't perfected, so also in the intercourse of the male and the female there is need of due measure.
And on account of this it happens to many [men] and many [women]
that they don't generate with one another, but that after being separ-   25
ated, they do generate, and these oppositions happen sometimes to
young [men and women] and sometimes to older [ones], likewise
with respect to coming into being and sterility, and generating males
and generating females. And also region differs from region in these
[matters], and water from water, on account of the same causes; for
the nutriment, and the condition of the body, come to be of a certain   30
sort especially on account of the blend of the surrounding air and of

---

8   The "turnings" at 767a7 would be the moon's maximum distances to the
    north and south of the equator, corresponding to the summer and winter turn-
    ing points, i.e., solstices, in the sun's yearly motion.

the [foods] that enter [the body], and especially on account of the nutriment that consists of water; for this [is what] they take in most of all, and in all [foods] there is this [as] nutriment, even in dry [foods]. Wherefore also hard and cold waters in some cases produce

35   sterility and in others the bringing forth of females.

Chapter 3

The same causes [are] also [responsible] for [it being the case] that some [offspring] come into being resembling those who produced [them] while others don't resemble [them], and [that] some [re-

767b   semble the] father and others [the] mother with respect to the body as a whole as well as with respect to each part, and [that they re-semble] them more than [they do] their earlier ancestors, and these [more] than any chance [beings], and [that] the males [resemble] the father more and the females the mother, and [that] some [re-semble] no one among their kindred but nevertheless [do resemble]

5   some human being, at any rate, while some don't even [resemble] a human being in form but already [resemble] a monstrosity. And in fact the one who doesn't resemble his parents is already in a certain manner a monstrosity; for nature in these has gone outside the kind in a certain manner. But the first beginning [of all this is] that there come into being a female and not a male; but this [beginning] is necessary for nature. For the kind [consisting] of those that are

10   separated into the female and the male needs to be preserved, and since [it is] possible for the male sometimes not to gain the mastery, either on account of youth or old age or on account of some other such cause, [it is] a necessity that there come to be a bringing forth of females among the animals. But the monstrosity [is] not neces-sary with a view to the cause for the sake of something, i.e., that

15   [consisting] of the end, but as a concomitant [it is] necessary, since one has to grasp its ruling beginning, at any rate, from there.[9] For when the spermatic residue in the menses is well concocted, the motion of the male will produce the shape in accord with itself; for

---

9   It is perhaps worth noting that the Greek word *teras*, translated here as "mon-strosity," originally meant a sign or portent sent by the gods or a monster of divine origin.

to say semen or the motion that causes each of the parts to grow
doesn't make any difference, nor [to say] the [motion] that causes
growth or the one that constitutes from the beginning; for [there   20
is] the same measure (*logos*) of the motion. So that if it gains the
mastery, it[10] will produce a male and not a female, and [one] that
resembles the generator and not the mother; but if it hasn't gained
the mastery, in accord with whatever power it hasn't gained the
mastery, it produces the deficiency that corresponds to it. I mean
"each power" in this manner; that which generates is not only a
male but a male of such a sort, for example, Koriskos or Socrates,   25
and not only is he Koriskos but also a human being. And in this
manner some [designations] belong more closely and others more
remotely to the one generating insofar as he is capable of generating, but not as [mere] concomitants, as for example, if the one generating is literate or is someone's neighbor. And that which is
distinctive and the particular always prevails more with respect to   30
coming into being. For Koriskos is both a human being and an animal, but his [being] human is closer to what is distinctive than his
[being] an animal. And both the particular and the kind generate,
but more so the particular; for this [is] its being. For indeed that
which comes into being becomes also of a certain sort, but a certain   35
this,[11] and this is its being. Wherefore the motions from the powers
belonging to all such [designations] are present in the seeds, and
potentially even [from those] belonging to the earlier ancestors,
but more [from those] belonging to what in every case is closer to
some one of the particulars; I mean by "particular" Koriskos and   768a
Socrates. And since everything deviates not into any chance [thing]

---

10 At 767b21, the grammar of the sentence raises the question of whether what
    produces a male in this case is identical with the motion of or from the male
    generator, since the Greek word translated as "if it gains the mastery" is a gen-
    itive absolute, a construction normally not used to refer to the subject of the
    main clause. Also, in the second half of the sentence, the Greek word translated
    as "if it hasn't gained the mastery" is neuter in form, unlike the words for mo-
    tion or for semen (or for measure), so that it isn't entirely clear what the "it"
    is that Aristotle has in mind here.

11 At 767b34, Drossaart Lulofs adds the word *kai*, meaning "also," before the
    words translated here as "a certain this" (*tode ti*). In this he has the authority of
    one of the manuscripts, but a relatively minor one. The other manuscripts all
    omit the word, and I accept their reading, since Aristotle is not saying here what
    the being that is of a certain sort "also" is, but what it is that is also of that sort.

but into its opposite, [it is] also necessary for that which is not mastered in its coming into being to deviate and to become the oppo-
5   site in respect to the power according to which that which generates
and sets in motion didn't master [it]. If, then, it deviates insofar as
[the generator is] male, it becomes a female, whereas [if it deviates
insofar as he is] Koriskos or Socrates, it comes into being not resembling its father but its mother; for just as mother is opposed to
father in general, so also the particular [female] that generates [is
opposed] to the particular [male] that generates. And likewise also
10   in respect to the powers that come next; for it always changes into
the more closely connected of the earlier ancestors, both in the line
of the fathers and in that of the mothers. And some of the motions
are in [the seeds] actually, but others potentially, actually those of
the [male] that generates and of the universal, for example human
being and animal, but potentially those of the female and of the
15   earlier ancestors. So then in deviating it changes into its opposites,
but the motions that are doing the fashioning weaken into those
close [to them], for example, if the motion of the generating [male]
is weakened, it changes by the least difference into that of his father, and secondly into that of his grandfather; and in this manner,
then, both in the case of the males and in that of the females,[12] the
20   [motion] of the generating female [changes] into that of her
mother, and if not into this, into that of her grandmother; and likewise also in the case of the more remote [ancestors]. Now then it
is most natural [for the motions from the male parent], insofar as
[he is] male and insofar as [he is the] father, to master or to be mastered at the same time; for the difference is small so that [it is] not
hard for both to happen at the same time; for Socrates is a man
[i.e., a male human], a certain one of such a sort. Wherefore for
25   the most part males resemble their father and females their mother,
for the deviation into both came about at the same time. The female
is opposed to the male and the mother to the father, and the deviation [is] into the opposites. But if the motion from the male gains

12 At 768a18–19, Drossaart Lulofs would delete the words translated here as
"both in the case of the males," though without any ancient authority. A translation of the passage, if one were to accept his reading, would be "and in this
manner, then, also in the case of the females, ..." But the manuscript reading
seems defensible, with the males and the females in question being the male
and female offspring of the generating female.

the mastery but that from Socrates doesn't gain the mastery, or
[if] this one does but that one doesn't, then it turns out that there   30
come into being males resembling their mother and females [re-
sembling] their father. And if the motions are weakened, and it remains insofar as [it is] male, but the one from Socrates is weakened
into that from his father, there will be, in accord with this argument, a male resembling its grandfather or one of the other, more
remote ancestors; but if it is mastered insofar as [it is] male, there
will be a female and, for the most part, [one] resembling the   35
mother, but if this motion is also weakened, the resemblance will
be to the mother's mother or to one of her more remote [ancestors],   768b
in accord with the same argument. And [there is] the same manner
[of explanation] in the case of the parts as well; for of the parts as
well some often resemble the father, and others the mother, and
others some of the earlier ancestors; for motions of the parts as
well are in [the seeds], some actually and others potentially, as has   5
often been said. But one has to make [some] assumptions in general, one the one [just] stated, that some of the motions are in [the
seeds] potentially and others actually, and two others, that if mastered it deviates into its opposite, and that if weakened [it changes]
into the next motion, if weakened less, into the one nearer, and if
more, into the more remote one. And in the end they are so con-   10
fused together that they resemble no one among their relatives and
kindred, but there is left only what is common, i.e., to be a human
being. And responsible for this [is] that this accompanies all the
particulars; for human being is universal, but Socrates [the] father,
and the mother, whoever she may be, [are] among the particulars.   15
Responsible for the motions being weakened, on the one hand, [is]
that that which acts is also acted upon by that which is acted upon
[by it], for example, that which cuts is blunted by that which is cut
and that which heats is cooled by that which is heated, and in general that which sets in motion, apart from the first [mover], is
moved in return with some motion, for example, that which pushes
is pushed in return somehow and that which squeezes is squeezed   20
in return, and sometimes it is even acted upon, in general, more
than it acts, and that which heats is cooled and that which cools is
heated, sometimes having acted to no effect and sometimes to less
of one than it underwent. These [things] have been spoken about
among the [things] that have been determined concerning acting

and being acted upon, in which sorts of the beings acting and being
25   acted upon are present.[13] On the other hand, that which is acted
upon deviates and is not mastered, either on account of a deficiency in the power of that which concocts and sets in motion or
on account of the large amount and coldness of that which is being
concocted and given determinate [shape]; for by mastering at one
point but not mastering at another, it makes that which is being
constituted manifold in shape, as happens in the case of athletes
30   because of their eating a lot; for since nature, on account of the
large amount of nutriment, is not able to gain the mastery so as to
produce proportional growth and [so that] the shape remain similar
[throughout], the parts come into being of different sorts and
sometimes in such a way that pretty much nothing resembles what
it was before. Nearly the same as this [is] the sickness called saty-
riasis; for in this, on account of the large amount of unconcocted
35   discharge or *pneuma*[14] that finds its way to parts of the face, the
769a   face appears [to be that] of another animal, i.e., of a satyr. And so
on account of what cause females and males come into being, and
some [come into being] resembling their parents, females [resem-
bling] females and males males, while contrariwise some females
[resemble] their father and males their mother, and in general some
5   resemble their ancestors whereas others don't [come into being re-
sembling] any [of them], and these [things] both in respect to the
body as a whole and to each of the parts, it has been determined
concerning all [these things]. But some of those who speak about
nature, and others,[15] have spoken concerning these [matters], on
account of what cause [offspring] come into being like and unlike
their parents. Now they state two manners of cause. For some as-
10   sert that from whichever of the two [parents] more seed comes,

---

13 Cf. *On Coming into Being and Perishing* 324a24–b13.

14 At 768b35. Drossaart Lulofs would delete the words translated as "or *pneuma*,"
even though they appear in all the manuscripts.

15 At 769a7, I follow Drossaart Lulofs, who reads *heteroi*, in keeping with the great
preponderance of the ancient authorities. One manuscript, however, reads *het-
eron ti*, which would lead to a translation of the beginning of this sentence as
"But some of those who speak about nature have said also something other concerning these [matters], ..."; also, some editors accept the modern emendation
hetera, which would lead to the translation, "But some of those who speak about
nature have said also other [things] concerning these [matters], ..."

[the offspring] comes into being resembling this one more, likewise
all [the body resembling] all and part [resembling] part, on the assumption that seed comes from each of the parts; and if an equal
[amount] comes from each of these two, [they assert] that [the off-
spring] comes into being similar to neither. But if this is false and
[seed] doesn't come from all [the body], [it is] clear that what was
said [by them] would not be responsible for the likeness and un-   15
likeness. Further, they are not easily able to determine how [there
comes into being] a female that at the same time resembles her father or a male that resembles his mother; for those who speak as
Empedocles or Democritus [do] say [things that are] impossible
in another manner concerning the cause of the female and male;
and those who [say that] by more or less [seed] coming from the   20
male or female, [that] on account of this the one comes into being
female and the other male, wouldn't be able to show in what manner there will be the female resembling the father or the male the
mother; for [it is] impossible for more [seed] to come from both at
the same time. Further, on account of what cause does [the off-
spring] come into being resembling for the most part its earlier an-   25
cestors, even those that are distant? For none of the seed came from
these, at any rate. But those who speak about likeness in the manner
that remains speak better, both in the other [things they say] and
in this. For there are some who assert that the semen, while being
one, is as it were a kind of seed-aggregate of many [seeds]; so just
as if someone were to blend many juices into one liquid and then   30
take [some away] from it, he would be able in every case to take it
unequally from each juice, but sometimes more of this sort and
sometimes more of that sort, and sometimes to take [some] of this
but to take none of that, [they say that] this happens[16] also in the
case of semen, which is mixed from many [seeds]; for the one of
the generators from which most [seed] comes to be in [the off-   35
spring,] [they say that] it comes into being resembling this one in
shape. And this argument, while it is unclear and fabricated in

16 At 769a33, I follow Drossaart Lulofs in accepting the modern emendation
   sumbainein. The manuscript reading is *sumbainei*, according to which the
   translation would omit the bracketed words "[they say that]." One reason for
   accepting this emendation is that the bracketed words "[they say that]" later
   in the sentence are required by the reading of the manuscripts.

769b   many ways, wishes to say [something] better, that what it calls a
seed-aggregate is present not actually but potentially; for in the
former way [it is] impossible, but in the latter possible. But [it is]
not in fact easy, giving one manner of cause, to state the causes
concerning all [the phenomena], [that] of there coming into being
5   female and male, and [the cause] on account of which the female
[is] often similar[17] to the father and the male to the mother, and
again of likeness to the earlier ancestors, and further, on account
of what cause there sometimes [comes into being] a human being,
but [one] not much like any of these, and sometimes, proceeding
in this way, in the end that which comes into being doesn't even
10   appear [to be] a human being, but only some animal, and these are
in fact called monstrosities. And indeed [what comes] next after
the [things] that have been said is to speak about the causes of such
[things]. For in the end, if the motions are weakened and the
material is not mastered, there remains what is most universal, and
this is the animal. And [people] assert that the [child] that comes
into being has the head of a ram or of an ox and in the other [kinds]
15   likewise [that] of a different animal, a calf the head of a child or a
sheep [that of] an ox. But while all these [things] happen on account of the causes that have been stated earlier, there are [in fact]
none of the [beings] that they say, but only [some] resembling
[them], which very [thing] comes about even among those that are
not deformed. Wherefore mockers often liken some of those who
20   are ugly to a goat breathing fire and others to a butting ram. And
a certain physiognomist reduced all faces to [those of] two or three
animals, and he often persuaded [people] when he spoke. But that
it is impossible for a monstrosity of that sort to come into being,
one animal in a different one, is made clear by the times of gestation, which are far different [in the case] of a human being and of
25   a sheep and of a dog and of an ox; [it is] impossible for each to
come into being except in its own proper times. So then some of
the monstrosities are spoken of in this manner, but others by their

---

17   At 769b6 I follow Drossaart Lulofs, who adds within angular brackets the
word *homoion*, which I translate as "similar," even though it is not present in
the manuscripts. It was apparently present, however, in the manuscript used
as the basis for the Arabic translation. And if it were not included in the text,
something like it would have to be understood.

having a shape with multiple parts, since they come into being with many feet and many heads. And the accounts of the cause, those concerning monstrosities, are nearly alike, and about the same in a certain manner as those concerning deformed animals. For the   30
monstrosity is a sort of deformity.

Chapter 4

Now Democritus asserted that monstrosities come into being on account of two semens falling [into the uterus], the one having set forth earlier and the other later, this latter one too, after having left [the male],[18] so that the parts grow together and become intermixed. And since it happens in birds that copulation always takes place quickly, he asserts that the eggs and their   35
color become intermixed. But if it happens that from one seed, i.e., one act of intercourse, more [offspring] than one come into being,   770a
which very [thing] is manifest, [it is] better not to go around in a circle, disregarding the shortest way; for in such cases especially [it is] necessary for this [phenomenon] to happen, whenever the seeds are not separated but come [into the uterus] at the same time. And so if one has to hold the semen from the male responsible, one would   5
have to speak in this manner; but one ought to suppose that in general the cause [is] more in the material and in the embryos that are being constituted. Wherefore also monstrosities of this sort come into being quite rarely among those that bring forth one [offspring at a time], more among those that bring forth many, and most of all among birds, and of birds among hens; for these bring forth many   10
not only by laying [eggs] often, as does the kind consisting of pi-

---

18 At 769b32–33, Drossaart Lulofs would emend the text in two ways, by adding the words *kai mē exelthousan* (at b32) after the word that I have translated as "having set forth earlier" and by deleting in the next line the word (*exelthousan*) that I have translated as "after having left [the male]." Neither of these changes has any manuscript support, although the first one is in keeping with the old Arabic translation, as well as with a corrected version of William of Moerbeke's Latin translation. According to Drossaart Lulofs' text, the relevant portion of this sentence would be translated as "the one having set forth earlier, and not having left [the uterus], and the other later, this latter one too coming into the uterus...." In any case, the sense of the passage seems clear enough.

geons, but also by carrying many embryos at the same time and by copulating at every season. Wherefore they also lay many double

15 [eggs]; for the embryos grow together on account of being near one another, just as many of the fruits sometimes [do]. For from as many of these [eggs] as whose yolks are separated by the membrane, there come into being two separate chicks that have nothing extraordinary; but from as many as whose [yolks are] continuous and nothing keeps them apart, from these there come into being chicks that are

20 monstrous, having [one] body and one head, but four legs and [four] wings, on account of the upper [parts] coming into being out of the white and [doing so] earlier, with the nutriment for them being drawn from the yolk, whereas the lower part comes into being later, and its nutriment is one and without distinction. There has also been seen a snake with two heads on account of the same cause;

25 for this kind too lays eggs and lays many. But the monstrous is rarer in their case because of the shape of their uterus; for because of its length the multitude of eggs lie in a row. And with regard to bees and wasps nothing of this sort comes about; for their offspring are

30 in separated cells. But with regard to hens the opposite happens, by which [it is] also clear that one ought to hold the cause of such [things to be] in the material; for among the others as well [they happen] more among those who bring forth many [at a time]. Wherefore [they happen] less in [the] human being; for in most cases it brings forth one and brings it forth perfected, since even among these, in the places in which the women give birth to more than one [at a time] this happens more, for example in Egypt. But

35 than one [at a time] this happens more, for example in Egypt. But in goats and sheep it comes about more; for they bring forth more [at a time]. And further, [it comes about] more among those with

770b multiple toes; for those sorts of animals bring forth more than one [at a time] and bring forth [offspring that are] not [yet] perfected, like the dog; for most of these bring forth [their offspring] blind; [the] cause on account of which this happens and [the] cause on account of which they bring forth many must be stated later. But the way has been prepared for bringing forth monstrosities by their nature, by their not generating [offspring] like [themselves] on ac-

5 count of their imperfection [at birth]. And the monstrosity too belongs to [the class of] those that are unlike. Wherefore there is an overlap between this accident and the [animals] of this sort in their nature. For instance, [it is] in these most of all [that] there also come

162

into being the so-called *metakhoira*.[19] And the condition of these is in a certain respect monstrous; for to fall short and to be in addition [are] something monstrous. For the monstrosity is [one] of the [things] contrary to nature in a certain respect, but not contrary to all nature but to that which is for the most part; for contrary to that which is always and that which is from necessity,[20] nothing comes into being contrary to nature, but [there is coming into being contrary to nature] among the [things] coming into being for the most part in this way but that admit of also [happening] otherwise, although even in as many of these [things, i.e., animal births,] as happen contrary to this order, yet always not by chance, [what happens] is thought to be less a monstrosity on account of the fact that even

10

15

---

19 Cf. 749a1–6, together with Book Two, n. 71. Here Aristotle is probably thinking of the fact that the *metakhoira* are smaller at birth than normal piglets (cf. *On the History of Animals* 573b4–5).

20 At 770b11, Drossaart Lulofs, like all the other modern editors I am aware of, reads the preposition *peri* ("as regards"), which occurs for the first time in a printed Renaissance text of Aristotle, instead of *para* ("contrary to"), the reading of all the manuscripts and of the other older authorities. According to Drossaart Lulofs' text, the beginning of this clause would be translated, instead of as "for contrary to that which is always and that which is from necessity," as "for as regards that which is always and that which is from necessity." Since the word that I have translated here as "that" is in both cases the feminine definite article, and since the context suggests that the noun to be supplied with them is *phusin*, the feminine word for "nature," editors seem to fear that the reading *para*, and hence the translation "contrary to the [nature] which is always and the [nature] which is from necessity," would make the words "contrary to nature" in the continuation of the clause ("nothing comes into being contrary to nature") superfluous and tautological. Yet Aristotle's decision not to say explicitly, in the first part of this clause, "the nature which is always and the nature which is from necessity" might instead be based on the thought that what is always and what is from necessity in the strictest sense of these words is not nature, or the order of the world within which we live (cf. *On the Parts of Animals* 641b23–26; *On the Heaven* 301a4–11; and *Physics* 252b26–28); but "being" in a wider sense, a Greek word for which is the feminine noun *ousia*. Being, in this sense, though the source and substrate of our natural order, does not have an order itself, but is simply always what it is. And if this is what Aristotle has in mind, *ousian* (being), not *phusin* (nature), would be the noun to be supplied with the two definite articles; and the clause as a whole, without emendation, might well be translated as, "for contrary to the [being] that is always and the [being] that is from necessity, nothing contrary to nature comes into being." This claim would be true because nothing ever happens contrary to being in this widest sense and, in particular, nothing that happens contrary to nature, as some things do, is contrary to it.

what is contrary to nature is in accord with nature in a certain manner, whenever the nature in accord with the form doesn't master
that in accord with the material. Wherefore neither do they speak
of such [things as] monstrosities, nor in the other [cases], in as
many as in which something customarily comes into being, as in
20   fruits. For there is a certain vine which some call smoky, which if
it bears black bunches of grapes they do not judge [it to be] a monstrosity on account of its customarily doing this very often. Responsible [for this is] that it is between white and black in its
nature, so that the change is not from a distance and not, as it were,
contrary to nature; for [it is] not into another nature. And these
25   [things] happen among those [animals] that bring forth many on
account of the many [embryonic] offspring hindering one another's
completions and the generative motions. Now someone might be
perplexed about the bringing forth of many and having more [than
the normal number] of parts, and [about] the bringing forth of few
30   and that of one and the lack of [some] parts. For sometimes there
come into being some [animals] that have too many toes and others
[that have] one alone, and with regard to the other parts [it is] in
the same manner; for they are both in excess and come into being
curtailed [in number]. And some [animals come into being] even
having two genital organs, one of a male and the other of a female,
35   both among human beings and especially as regards goats; for there
come into being those that they call *tragainae* [i.e., hermaphroditic
goats], on account of their having a genital organ of a female and
of a male; also, a goat has in fact come into being that had a horn
771a on its leg. Changes and deformations[21] come into being also with
regard to the internal parts, by [animals] either not having some
or [having them] curtailed [in number] or too many or with their
places changed. Now no animal has ever come into being not having a heart, but [there have been animals] not having a spleen and
having two, and [having] one kidney. No [animal has come into
5   being] not having a liver, but not having a whole [one]. And all
these [things have happened] among those that have been per-

---

21 At 771a1, Drossaart Lulofs would add the words *kai pleonasmoi*, meaning "and
excesses [in number]" after the words translated here as "and deformations."
He has support for this addition from the Arabic translation, but the words
don't appear in any of the manuscripts.

fected and that live. [Animals] are found also not having a gall bladder, though being of a nature [to have one]; and some [are found]
having more than one. And [these parts] have also come into being
with their places changed, the liver on the left and the spleen on
the right. And these [things] have been seen in the animals that   10
have been perfected, as has been said; but in those that are [merely]
brought forth,[22] [things have been seen] that have much and all
sorts of confusion. Now some [animals], though departing a little
from their nature, usually live, but others, [departing] more, [usu-
ally] don't live, whenever that which is contrary to nature comes
to be in the [parts] that are sovereign for living. Now the inquiry
concerning these [things] is whether one ought to hold [that there   15
is] the same cause of bringing forth one and of the lack of parts,
and of excess [in the number of parts] and of bringing forth many,
or [that the cause is] not the same. And in the first place, on account
of what [cause] some [animals] bring forth many while others bring
forth one, this [is something that] someone might believe is reasonably wondered at. For the biggest of the animals bring forth
one, for example, elephant, camel, horse, and the solid-hoofed   20
[animals]; and some of these [are] bigger than the other [animals],
while others are far superior in size. But dog and wolf and the [ani-
mals] with multiple toes pretty much all bring forth many, even
the small [ones] among those of this sort, for example, the class of
mice. And the cloven-hoofed [animals] bring forth few, except the
pig; this is among those that bring forth many. [Now these facts
are wondered at], for [it is] reasonable for the big [animals] to be   25
able to generate more [offspring] and to bear more seed. But this
very [thing] that is wondered at [is] responsible for [our] not won-
dering; for on account of their size, they don't bring forth many;
for the nutriment in those of this sort is used up for the growth of
their body, but in those that are smaller, nature, having taken away
from their size, adds the excess to their spermatic residue. Further,   30
[it is] necessary for the generating seed of the bigger [animal] to

22 At 771a11, the word translated as "that are [merely] brought forth" (*tik-*
    tomenois) appears in only one rather minor manuscript, with possible support
    from several ancient translations. I accept this reading, along with Drossaart
    Lulofs, even though the preponderance of the manuscripts read *eirēmenois*,
    meaning "that have been mentioned."

be of a larger amount, and for that of the smaller ones [to be] small [in amount]. And many small [embryos] could come into being in the same [place], but for many big ones [it is hard]. And to those of 35 medium sizes nature has given the medium [number of offspring]. Now then we have stated earlier the cause of some of the animals 771b being big, some smaller, and some medium-sized.[23] [Some] of the animals bring forth one, some bring forth a few, and some bring forth many. And for the most part, the single-hoofed [animals] bring forth one, the cloven-hoofed bring forth few, and the many-toed bring forth many. Responsible for this [is] that for the most part 5 the [animals'] sizes are distinguished in accord with these differences [with regard to feet]. However, it is not like this in all cases; for bigness and smallness of bodies [is] responsible for bringing forth few and bringing forth many, not that the kind is single-hoofed or many-toed or cloven-hoofed. And [here is] evidence of this; for the elephant [is] the biggest of the animals, and it is many-10 toed, and the camel, though it is the biggest of the remaining ones, [is] cloven-hoofed. And not only among the land animals, but also among those that fly and among those that swim, the big ones bring forth few and the small ones bring forth many, on account of the same cause. Likewise also in the case of the plants, [it is] not the biggest ones [that] bear the most fruit. And so it has been said on 15 account of what [it is that] some of the animals bring forth many, some bring forth few, and some bring forth one, but within the perplexity that has just now been mentioned, someone might reasonably wonder rather about those that bring forth many, since the animals of this sort manifestly become pregnant from one act of copulation. And the seed of the male—whether it contributes to 20 the material of the embryo by becoming a part [of it] and by mixing with the seed of the female, or whether in fact [it is] not in this manner but, as we assert, by bringing together and fashioning the material in the female and the spermatic residue, as fig juice does the liquid of milk—on account of whatever cause does it not bring 25 to perfection one animal that is big (as in this case the fig juice isn't separated by giving shape to a certain amount, but the more [milk] that it enters into and the more [of it there is], so much the greater is what is solidified)? Now to assert that the regions of the uterus

---

23 This is perhaps a reference to *On the Parts of Animals* 689b29–32.

draw in the seed and that it becomes multiple on account of this, on account of the multitude of the regions and the cotyledons[24] not being one, is [to say] nothing; for often two [embryos] come into being in the same region of the uterus, and in those that bring   30 forth many, when [the uterus] is full of embryos, they manifestly lie in a row. This [is] clear from the *Dissections*. But just as for each of the animals as they are being perfected there is a certain size, both in the direction of the bigger and that of the smaller, bigger or smaller than which they couldn't become, but in the intermedi-   35 ate interval of size they acquire their excess and deficiency in rela- tion to one another, and a human being, or any one whatsoever of   772a the other animals, becomes bigger and another smaller, so also the spermatic material out of which [an embryo] comes into being is not indeterminate, neither in the direction of the bigger nor that of the smaller, so as [for that] to come into being from any amount [of it] whatsoever. In as many of the animals, then,[25] as on account of the cause stated release more residue than [the fitting amount]   5 for [the] ruling beginning of one animal, it is not possible for one to come into being out of all this, but [rather] so many as [the num- ber of which] has been determined by the fitting sizes, nor will the seed of the male, or the power in the seed, give form to any that is more or less than the natural [amount]. And likewise, if the male emits more seed, or more powers in the seed as it is being divided,   10 the greatest amount will not make anything bigger [than the fitting size], but to the contrary will destroy [the spermatic material] by drying it up. For neither does fire heat water more in proportion as there is more [of it], but there is a certain limit of the heat [in water], which when it is present, if someone increases the fire, it   15 no longer becomes hotter, but rather evaporates and finally disap- pears and becomes dry. And since the residue of the female and that from the male (in as many of the males as release seed) appear to need a certain proportionateness toward one another, in those of the animals that bring forth many, the male straightaway emits   20

24 Cf. 745b33; and Book Two, n. 57.

25 At 772a4, I read *oun*, with the great preponderance of the manuscripts, rather than Drossaart Lulofs' *goun*, which appears in only one relatively late one. According to Drossaart Lulofs' text, instead of the word "then," the transla- tion would read "at any rate."

[an amount] that is able when divided up to give form to more [than one], and the female [releases] so much as [is able] to become multiple formations. And the example stated regarding milk is not similar. For the heat of the seed does not constitute [something that is] only of such a size but [also] of a certain sort, but that in 25 the fig juice and the rennet only what is of such a size. So then in those [animals] that bring forth many, responsible for many embryos coming into being and not a single continuous [one] out of [them] all [is] this very [thing], that an embryo doesn't come into being out of however much [residue] it may be, but if there is only a little there will not be [one], nor if there is too much; for the power both of that which is acted upon and of the heat that acts 30 upon [it] is determinate. Likewise also in those of the animals that bring forth one and [that are] big, there don't come into being many [embryos] from much residue; for also in [the case of] those, out of a certain amount [of it] what is produced is of a certain size. And so they don't release more material of that sort on account of the cause that was stated earlier; and that which they release is as 35 much, in accord with nature, as from which only one embryo comes into being. But if ever there comes more [of it], then they bring forth two. Wherefore such [embryos] even seem rather to be monstrous, because they come into being contrary to what is for 772b the most part and what is usual. But the human being has a share in all the classes; for it brings forth one and it brings forth few and it sometimes brings forth many, but as regards its nature it most of all brings forth one: on account of the wetness and heat of its 5 body it brings forth many (for the nature of seed is wet and hot), but on account of its size it brings forth few and brings forth one. And on account of this there results also that for it alone of the animals the times of gestation are irregular. For the others the time is one, but for human beings more than one; for they are born at seven [lunar] months and at ten months and at the times in be- 10 tween; indeed even those [born] at eight months live, though less often. And someone might understand what is responsible [for these facts] from the [things] that have just now been said, and it has been spoken about them in the *Problems*.[26] And concerning these [things] let it have been determined in this manner. For parts

---

26 There is no such discussion in the extant portions of the *Problems*.

being multiplied contrary to nature, the same [thing is] responsible as for the bringing forth of twins. For that which is responsible occurs already in the embryos if more material is given form than [is] 15 in accord with the nature of the part; for then [what] happens [is either] that they have a part bigger than the others, for example, a finger or a hand or a foot or one of the other extremities or limbs, or, if the embryo is split, that more [of these parts] come into being, just like eddies in rivers; for in these also, if the water that is carried along and that has a [certain] motion strikes against [something], 20 there come into being two formations, out of [the] one, that have the same motion; and it happens in the same manner in the case of embryos. And [the parts of the same form] are generally attached close to one another, but sometimes even far away, on account of the motion that comes to be in the embryo, and especially on account of the excess of the material returning to where it was taken away from, but having the form from [the place] where it arose as 25 an excess. In as many [animals] as turn out such as to have two genital organs, one of a male and the other of a female, one of the multiplied [parts] comes into being with the capacity to function and the other without it, because of its always being weakened in regard to [getting] nutriment inasmuch as it is contrary to nature, but it is attached just like tumors; for these too get nutriment even 30 though they are of a later origin and contrary to nature. If the fashioning [agent] gains the mastery, or if it is wholly mastered, there come into being two [of these organs] that are similar. But if in a way it gains the mastery but in a way is mastered, one [of them is] female and the other male; for it doesn't make any difference to state this, [namely, the] cause on account of which one comes into being female and the other male, with regard to the parts or with regard to the whole. And as for as many [animals] as come into 35 being lacking parts of this sort, for example, some extremity or [one] of the other limbs, one ought to hold [that there is] the same cause as if what is coming into being wholly miscarries; and there 773a come to be many miscarriages of embryos. While outgrowths differ from the bringing forth of many [offspring] in the manner that has been stated, monstrosities [differ] from these by most of them being a growing together [of embryos]. But some [occur] also in this manner, if they come about in the case of bigger and more sovereign parts, for example, some have two spleens and more [than 5

two] kidneys. Further, there are instances of the parts changing position, when their motions are diverted and their material changes position. One ought to hold the animal that is monstrous to be one, or many that have grown together, in accord with the
10   ruling beginning, for example, if the heart is such a part, [one ought to hold] that which has one heart [to be] one animal, and the excessive parts outgrowths, and [one ought to hold] those that have more than one [heart] to be two, and that they have grown together on account of the conjoining of the embryos. It often happens to many, even among the animals that don't seem to be deformed, that after having already been perfected, some of their channels
15   are closed up and others are diverted. And in fact in some females the mouth of the uterus has remained closed, and when it was already the time of the menses and pains were coming upon [them], in some it burst open spontaneously and in others it was cut open by doctors; and it came to pass that some perished, if the rupture
20   either came about with violence or wasn't able to come about. And in some boys the end of the penis hasn't coincided with the channel through which the residue from the bladder passes but [the channel] was underneath it; wherefore they urinate sitting down, and when their testicles have been drawn up they seem to those at a distance to have a genital organ of a female and of a male at the
25   same time. And in the case of some animals, both sheep and others, the channel for [the residue of] dry nutriment has come into being closed up, and in fact a cow was born in Perinthos in which finely sifted nutriment went through the bladder, and when its anus was cut open it quickly grew back together, and they didn't have the
30   power to keep it open. And so concerning the bringing forth of few and the bringing forth of many and concerning the outgrowth of parts that are in excess,[27] and further, concerning the [animals] that are monstrous, it has been spoken.

---

27   At 773a31, I follow Drossaart Lulofs in deleting the words *ē elleipontōn*, which in the manuscripts follow the word translated here as "in excess." If they were included, instead of "parts that are in excess," the translation would read "parts that are in excess or deficient." Although these words are present in our manuscripts, they were apparently lacking in the text used as the basis for the Arabic translation.

Chapter 5

Some among the animals don't become doubly pregnant[28] at all whereas others do become doubly pregnant, and of those that become doubly pregnant, some are able to bring the embryos to birth while others [are] sometimes [able to] but sometimes not. Responsible for [some] not becoming doubly pregnant [at all is] that they bring forth one [at a time]. For the solid-hoofed animals and those bigger than these don't become doubly pregnant; for on account of their size, their residue is used up on the [already existing] embryo. All of these are big of body, and the fetuses (*embrua*) of the big are correspondingly big as well; wherefore the fetus (*embruon*) of elephants is as big as a calf. But those that bring forth many [at a time] become doubly pregnant on account of the fact that even the [conceiving] of more than one is [an instance of] one embryo added[29] to another. And among these [animals], all those that are big, like the human being, if the one copulation comes about soon after the other, bring the subsequent embryo to birth; indeed, this sort of thing has in fact been seen to have happened. And [what is] responsible [is] what has been said; for even in the one act of intercourse the seed that is discharged is more [than enough for one embryo], and this, when divided up, produces the bringing forth of more than one, one of which comes later. But when the copulation happens to come about after the embryo has already grown, there is sometimes a double pregnancy, though rarely, on account of the uterus in women being closed up for the most part for as long as they are pregnant. But if it ever happens (and this has in fact come about), [the mother] isn't able to bring [the second embryo] to perfection, but embryos are cast out that are nearly the same as what are called

35

773b

5

10

15

---

28 At 773a33, the Greek word that I have translated as "become doubly pregnant" means more precisely to become pregnant a second time as the result of copulation while pregnant already. The beginning of this sentence is usually translated as something like, "In some animals superfetation doesn't occur at all...." But while "superfetation" is indeed a genuine English term for this phenomenon, I have avoided it on the grounds of its unfamiliarity.

29 At 773b7, the Greek word that I have translated as "embryo added" is *epikuēma*, which is usually translated as "superfetation," though in this clause Aristotle is referring, at least in part, to the more common phenomenon of producing more than one offspring as the result of a single act of intercourse.

abortions. For just as, in the case of the [animals] that bring forth one, on account of their size all the residue is directed to the [em-
20   bryo] that is present before, so also in these, except that in those [it happens] immediately, but in these, when the fetus (*to embruon*) has grown; for then they are in nearly the same condition as those that bring forth one. And likewise, on account of the human being being by nature [an animal] that brings forth many, and there being a certain surplus in the size of the uterus and [the amount] of residue,
25   but not so much as to bring a second [fetus] to birth, a woman and a mare, alone of the animals,[30] allow copulation while they are pregnant, the [woman] on account of the cause that has been stated, and the mare on account of the barrenness of its nature and there being a certain surplus size of its uterus, [it being] more than enough for the one [fetus], but less than so as to be doubly pregnant with another [that is] perfected. And it is by nature given to sexual activity on account of being in the same condition as those
30   [women] that are barren; for those are of that sort on account of there not being a purgation [of the menses] (and this is like ejaculation in males), and female horses discharge a minimal purgation. And in all the [kinds] that bring forth animals, the barren among the females are given to sexual activity on account of being in
35   nearly the same condition as males when their seed has been col-
774a lected but is not being eliminated. For in females the purgation of the menses is the evacuation of seed; for the menses are uncon-cocted seed, as has been said earlier.[31] Wherefore also, as many women as lack self-control in regard to intercourse of this sort cease from their excitement when they have brought forth many
5   [children]; for when the spermatic residue has been secreted, it no longer produces a desire for this intercourse. Among birds the fe-males are less given to sexual activity than the males, on account of their having the uterus [up] by the diaphragm, whereas the males [are] the opposite; for they have their testicles drawn up in-
10   side, so that if the class[32] of such birds is by nature productive of

---

30 In saying "alone of the animals," perhaps Aristotle means "alone of the animals that are big," but I am not certain.

31 726b30–727a2.

32 At 774a9, I read *to genos*, in accordance with the manuscripts, rather than *ti genos*, a modern emendation accepted by Drossaart Lulofs. Drossaart Lulofs'

[much] seed, it is always in need of this intercourse. And so in the females the uterus descending, and in the males the testicles being drawn up, are attributes that promote copulation. So then [the] cause on account of which some [animals] don't have double pregnancy at all, whereas some do become doubly pregnant, and some [that become doubly pregnant] sometimes bring the embryos to   15
birth and sometimes not, and on account of what cause some of such [animals] are given to sexual activity while others are not given to sexual activity, has been said. Some of the [animals] that become doubly pregnant are able to bring the embryos to birth even if the [second] copulation comes after an interval of much time, [namely, those] whose kind is productive of [much] seed and whose body is not big and that are among those that bring forth many [at a time];   20
for on account of bringing forth many they have room in the uterus, and on account of being productive of [much] seed they discharge much residue [by way] of the purgation; and on account of their body not being big but their residue being in excess, [being] in a greater measure than the nutriment [required] for the embryo, they are able to constitute animals even later and to bring   25
these to birth. Further, the uterus of such [animals] doesn't close up on account of there being a surplus of residue for purgation. And this has happened even in the case of women; for purgation comes about in some while they are pregnant, even up to the end. But in these [it is] against nature—wherefore it harms the embryo—whereas for the animals of this sort it is in accord with nature; for   30
their body is so constituted from the beginning, for example, the [class] of hares; for this animal becomes doubly pregnant; for it is not [one] of the big [animals], and it brings forth many (for it is many-toed, and the many-toed bring forth many) and [is] productive of [much] seed. Their hairiness makes [this] clear; for the amount of their hair is excessive; for this alone of the animals has hair under its   35
feet and inside its jaws. And its hairiness is a sign of a large amount of residue, wherefore also among human beings the hairy ones are   774b

---

text would be translated, instead of as "the class," as "any class." The emended text is perhaps easier to interpret (so long as it doesn't suggest a doubt as to whether there are any birds of this sort), but the manuscript reading is defensible if by "the class of such birds" Aristotle means "the class consisting of such birds [i.e., the males]."

given to sexual activity and are more abundant in seed than are the smooth. As for the hare, it often carries some of its embryos [in a still] unperfected [state] but delivers others of its young [already] perfected.

Chapter 6

5   Of those that bring forth animals, some deliver animals [that are] unperfected and some [deliver them] perfected, those that are solid-hoofed and those that are cloven-footed, perfected, and most of those that are many-toed, unperfected. Responsible for this [is] that those that are solid-hoofed bring forth one [at a time], while those that are cloven-footed either bring forth one or bring forth two, for the most part, and it is easy to nourish the few to perfec-
10   tion. And those of the many-toed that bring forth [their offspring] unperfected all bring forth many; wherefore, though they are able to nourish the embryos when they are newly formed, when they have grown and gained size, since the [mother's] body is unable to nourish them completely, they deliver them just like those of the animals that bring forth larvae. For of these too, some generate [off-
15   spring that are] pretty much unarticulated, like fox, bear, lion, and in nearly the same way some of the others as well; for pretty much all of them [are] blind [at birth], for example, these and, in addition, dog, wolf, jackal. Only the pig, while bringing forth many, brings them forth perfected, and this alone overlaps [the several classes]; for it brings forth many, like those that are many-toed, but it is cloven-footed and solid-hoofed; for there are surely solid-
20   hoofed pigs. Now it brings forth many on account of the nutriment for [increased] size being separated off into spermatic residue; for this [animal], for being solid-hoofed, doesn't have [great] size. But at the same time, and more [often]—as it were disputing with the nature of the solid-hoofed [animals]—it is cloven-footed. And on account of this it sometimes brings forth one and it brings forth two and for the most part it brings forth many, and it nourishes
25   them to perfection on account of the well-fed condition of its body; for it has sufficient and abundant nutriment, like rich earth for plants. Also some of the birds bring forth [offspring that are] un-perfected and blind, as many of them as bring forth many even

174

though they don't have bodies of great size, for example, crow, jay, sparrows, swallows, and as many of those that bring forth few as don't bring forth abundant nutriment together with their offspring, for example, ring-dove and turtle-dove and pigeon. And on account of this, if someone pokes out the eyes of swallows while they are still young, they are healed again; for [the eyes] are destroyed while the [birds] are coming into being but have not yet come into being, wherefore they grow and sprout from the beginning. And in general [birds] are [born] before becoming perfected on account of the [mother's] inability to nourish them completely, and they are born unperfected on account of being [born] early. This [is] clear also in the case of seven-month [children]; for on account of being unperfected, some of them often come into being without yet even having their channels, for example, [those] of the ears and of the nostrils, distinctly formed, but they become distinctly formed as [the children] are growing, and many of such [children] live. In human beings males come into being deformed more than females do, though in the other [animals] no more so. Responsible [for this is] that in human beings the male far surpasses the female in the heat of its nature, wherefore males, while in the womb, tend to be in motion more than females; and on account of being in motion they get broken more; for what is young [is] easily destroyed on account of its weakness. And also on account of this same cause females are not perfected in the same way as males, in [the case of] women.[33] For inside the mother the [human] female takes more time to be differentiated [into its parts] than the male, but after they have come forth, everything comes to perfection earlier in females than in males, for example, puberty, the prime of life, and old age; for females are weaker and colder in their nature, and one ought to take femaleness to be like a deformity [that is] natural. So then inside [the mother, the female] is differentiated slowly on account of its coldness (for the differentiation is concoction, and heat

30

35

775a

5

10

15

---

33 At 775a11, Drossaart Lulofs adds several clauses, which would be translated as "but in the other [animals it is] in the same way; for [in them] the female does not [develop] later than the male, as among women." Though these words have no manuscript authority, they derive some support from the Arabic translation and from a corrected version of the Latin translation by William of Moerbeke (while a later corrected version of this same translation suggests a somewhat different addition).

concocts, and what is hotter [is] easily concocted), but outside, on
20   account of its weakness, it quickly reaches its prime and its old age;
for all inferior [things] come to their perfection more quickly, as
in the works in accord with art, [so] also in the [beings] constituted
by nature. And also, on account of the cause mentioned, in human
beings those brought forth as twins are preserved less [if they are]
female and male, but in the other [animals these are preserved] no
25   less; for in the ones [it is] against nature for them to keep an equal
pace, given that their development doesn't come about in equal
times, but [it is] necessary for the male to be [born] late or for the
female to be [born] early, whereas in the other [animals, it is] not
contrary to nature. A difference is also found with regard to pregnancies between human beings and the other animals. For those
30   thrive more in their bodies for most of the time, whereas the majority of women feel discomfort in connection with pregnancy.
Now to some extent responsibility for these things is on account
of their manner of life. For since they are sedentary, they are more
full of residue, since in those nations in which the manner of life
of women is full of toil, pregnancy is not equally conspicuous, and
35   [women] accustomed to toil give birth easily, both there and everywhere; for toil uses up their residues, whereas for those who are
sedentary many such [residues] are present on account of their lack
of toil and their not having purgations while they are pregnant,
775b   and their labor is burdensome; whereas toil trains the breath so as
be able to hold it, [and] giving birth easily or with difficulty depends on this. So then, as has been said, these [things] contribute
to the difference of the condition between the other animals and
5   women, and especially the fact that in some of them there comes
to be little purgation and in others it is not noticeable at all, whereas
in women [there is] the most among the animals, so that when the
discharge doesn't come about on account of pregnancy, it gives
trouble to these; for even if they aren't pregnant, whenever the purgations don't come about, diseases occur; and [it is] at first after
10   having conceived [that] most women are troubled more; for the
embryo is able to prevent the purgations, and on account of its
smallness it doesn't use up a large amount of residue at first, but
later on, by taking a share of it, it eases [the trouble]; in the other
animals, on the other hand, because of its being a small amount,
[the residue] comes into being in a suitable proportion to the

176

growth of the fetuses (*embruōn*), and as the residues that hinder   15
nourishment are used up, [the mothers] are in better condition in
their bodies. And in aquatic [animals] and in birds it is in the same
manner. But in as many as no longer have a good condition of their
bodies at the point when their fetuses are becoming big, responsible
[for it is] that the growth of the fetus needs more nutriment than   20
that from the residue. And in some few among women, their bodies
happen to be in better condition while they are pregnant, and these
are all those in which the residues in their body are of a small
amount, so that it is used up along with the nutriment that goes to
the fetus (*embruon*).

Chapter 7

One must speak about that which is called a mole,[34] which though   25
it comes into being rarely in women, this condition does come into
being in some when they are pregnant. For they bring forth what
is called a mole. For it has in fact happened to a certain woman,
when she had had intercourse with her husband and thought she
had conceived, that though at first the bulk of her belly grew and
the other [things] came to pass in as reason would expect, when it   30
was time for giving birth, neither did she give birth nor did her
bulk become smaller, but she continued in this way for three or
four years until, after dysentery came about and she was in danger
from it, she brought forth [a mass of] flesh that they call a mole.
And in some [women] this condition grows old and dies with
them. Those of such [masses] that come out [of the body] become   35
so hard that they are cut in two [only] with difficulty, even with
iron. Now the cause of the condition has been spoken about in the
Problems;[35] for the embryo suffers the same [thing] in the womb   776a
as do those, among [meats] that are boiled, that are half-boiled,
and not on account of heat, as some assert, but rather on account
of the weakness of heat (for [the woman's] nature seems to lack

---

34 The Greek word translated as "mole" is *mulēs*, whose primary meaning is
    "mill" or "millstone." It refers to a mass of tissue, known today as a hydatid-
    iform mole, that forms inside the uterus at the beginning of a pregnancy.

35 There is no such discussion in the extant portions of the *Problems*.

power[36] and to be unable to bring [it] to perfection and to give a ter-
5   minus to its coming into being; wherefore it grows old with [the
woman] or lasts for a long time; for it has the nature [that it has] neither as [something] perfected nor as an entirely alien [growth]); for
lack of concoction is responsible for its hardness; for being half-
boiled is a sort of lack of concoction. [This] involves a perplexity as
to whatever [it is] on account of which [this condition] doesn't come
10   to be in the other animals, unless it has somehow entirely escaped
notice. But one ought to hold [what is] responsible [to be] that
woman alone of the animals has sufferings of the uterus and is excessive in regard to its purgations and is unable to concoct them;
thus, when the embryo is formed from a liquid that is poorly concocted, that which is called a mole then comes into being in women,
reasonably either most of all [in them] or [in them] alone.

Chapter 8

15   Milk comes into being in females, as many as produce animals inside
themselves, on the one hand [as something] useful for the time of
giving birth; for nature produced it in animals for the sake of [their
giving] nourishment externally, so that it should neither fall short at
all during this time nor be at all excessive; which indeed manifestly
20   happens if nothing comes about contrary to nature. Now in the other
animals, on account of there being a single length of time of their
gestation, its concoction coincides with this time; but in human beings, since there are multiple times, [it is] necessary [for it] to be
present at the first one; wherefore before seven months, milk is use-
25   less to women, and only then does it come into being, [being] useful.
And on the other hand, it happens reasonably also on account of the
cause from necessity [that it is] concocted at the final times; for at
first the secretion of residue of this sort is used up for the coming

---

36   At 776a3, the word translated as "to lack power" is *adunatein*, which appears in
only one of the ancient manuscripts. Drossaart Lulofs reads instead the word
asthenein, which would be translated as "to be weak," and this is roughly the
same as "to lack power," but his reading is supported only by the Arabic translation. The vast majority of the manuscripts read *dusthanatein*, which can mean
"to die with difficulty," but it doesn't make sense to me to say in this context
that "their nature seems to die with difficulty."

into being of the fetuses (*embruōn*); and in all [things] the nutriment
[is] what is sweetest and is concocted, so that when [what has] such
a capacity is taken away, [it is] necessary for what is left to become   30
salty and ill-flavored. But as the embryos become perfected, [there
is] more surplus residue (for what is being used up is less) and [it is]
sweeter, given that what is well concocted is not being taken away to
the same extent. For its expenditure doesn't come about any longer
for the molding of the fetus (*embruou*), but for a small amount of
growth, with the fetus (*embruon*) having as it were come to a halt at
that point on account of being at its end; for there is a sort of per-   35
fection even of an embryo. Wherefore it emerges [from the uterus]   776b
and changes its [manner of] coming into being, as having what (pl.)
belongs to it, and it no longer takes what (sing.) doesn't belong to it,
at which time milk comes into being, [being] useful. It collects in
the upper region, in the breasts, on account of the organization from
the beginning of the [body's] structure. For the [part] above the dia-   5
phragm is that which is sovereign over life, whereas the [part] below
is [the place] of nutriment and residue, so that as many of the animals
as are capable of locomotion, having self-sufficiency of nutriment
inside themselves, may change their places. And the spermatic residue is secreted from there [i.e., from the part above the diaphragm,]
on account of the cause that was stated in the speeches at the begin-   10
ning.[37] Now the residue of the males and the menses in the females
are of a bloody nature. And the ruling beginning of this and of the
blood vessels [is] the heart; and this [is] in these parts. Wherefore [it
is] necessary that the change of this sort of residue become noticeable
there first. Wherefore indeed the voices of both males and females   15
change when they begin to produce seed (for the ruling beginning
of the voice [is] from there; and it becomes of a different sort when
that which sets it in motion becomes of a different sort), and the
[parts] around the breasts are noticeably raised up even in males,
but more so in females; for on account of the excretion downwards   20
becoming great, the region of the breasts in them becomes empty
and spongy; and likewise in those [animals] that have their breasts
down below. Now both the [changed] voice and the parts around the
breasts become noticeable also in the other animals to those experi-
enced with regard to each kind, but the difference is greatest in

---

37 738b4–18; 747a19–20.

25   human beings. Responsible [for this is] that among females these females, and among males [these] males, have the most residue for
their size, in the former that of the menses, and in the latter the emission of seed.[38] So when the fetus (*embruon*) doesn't take in this sort
of residue, but prevents it from making its way outside, [it is] nec-
30   essary for all the residue to be gathered into the empty places, as
many as [are situated] on the same channels. And the region of the
breasts is of this sort in each [of the kinds that produce animals inside
themselves] on account of both causes, having come into being of
such a sort for the sake of the best as well as from necessity. And [it
is] here, at that time, [that] nutriment for the animals is constituted
35   and becomes concocted. Now it is possible to take the cause [just]
stated [to be responsible] for the concoction, but it is possible also
777a [to take] the opposite. For [it is] reasonable that the fetus (*embruon*),
being bigger, take in more nutriment so that there come to be less of
a surplus around this time; and what is less is concocted more
quickly. Now then that milk has the same nature as the secretion out
5   of which each [animal] comes into being [is] clear, and it has been
stated earlier.[39] For [it is] the same material that nourishes and out
of which nature constitutes the [animal's] coming into being. This
is the bloody liquid in the blooded [animals]; for   milk is blood
that has been concocted, not corrupted. And Empedocles either did-
n't understand correctly or he didn't make a good metaphor, [writ-
10   ing] that milk "comes into being on the tenth day of the eighth
month, a white pus."[40] For rottenness and concoction are opposites,
and pus is a sort of rottenness, whereas milk is among the [things]
concocted. And neither do the [menstrual] purgations come into
being according to nature in those who are suckling, nor do they conceive while they are suckling; and if they do conceive, their milk dries
15   up, on account of its being the same nature, [that] of milk and of
menses; and nature is unable to be so abundant as to suffice for both,
but if the secretion comes about in the one direction, [it is] necessary

---

38   At 776b27–28, Drossaart Lulofs would delete the words translated as "in the
former that of the menses, and in the latter the emission of seed," even though
they are present in all our manuscripts. These words were apparently not present, however, in the text used as the basis for the Arabic translation.

39   739b25–26.

40   Empedocles, fr. 68 (Diels/Kranz).

that it leave off in the other, unless it comes into being violently[41] and contrary to what is for the most part. And this is already [to say] contrary to nature; for in those [things] that are not impossible to be   20 otherwise but that admit [of it], what is according to nature is that which is for the most part. The coming into being of the animals has also been finely determined in regard to the times; for when the nutriment through the umbilical cord is no longer sufficient for the [fetus] carried in the womb on account of its size, at the same time[42] milk comes into being, [being] useful for the nourishing that comes to be [after birth], and since the nutriment is not going through the   25 umbilical cord, these blood vessels, around which the so-called umbilical cord is a covering, collapse, and on account of these [things, it is] also then [that] the going forth outside occurs.

Chapter 9

In all animals, the [way of] being born [that is] in accord with nature [is] head first, on account of the [parts] above the umbilical cord being bigger than those below. Thus [the parts], being suspended   30 from it as in a balance, incline toward the [greater] weight; and the bigger ones have more weight.

Chapter 10

The lengths of time of gestation for each of the animals are in fact determined for the most part in accord with their lifespans; indeed, [it is] reasonable that the comings into being of those that live longer

---

41 At 777a18, I read *ean mē gignētai biaion*, in accordance with the preponderance of the manuscripts. Drossaart Lulofs, however, reads *ean mē gignētai ti biaion*, a modern emendation with possible support from one relatively late manuscript. According to Drossaart Lulofs' text, instead of "unless it comes into being violently," the translation would be "unless something comes into being violently."

42 At 777a23, I follow Drossaart Lulofs in reading *hama*, a modern emendation that finds some apparent partial support from the Arabic translation. The manuscripts read *alla*, which would make the translation of the beginning of this clause "but milk comes into being," instead of "at the same time milk comes into being," and that would seem to leave the sentence without an appropriate main clause.

35   should take longer. Nonetheless, this is not responsible [for it], but
777b   this [is what] happens for the most part; for [to take another, similar,
case,] the bigger and more perfected of the blooded animals also
live for a long time, yet the bigger ones [are] not all longer-lived.
For the human being lives for the longest time of all, of those of
which we have credible experience, except for the elephant; but the
5   kind consisting of human beings is smaller than that of those with
long-haired tails and many others. Responsible for any animal whatsoever being long-lived [is] its being blended in nearly the same way
as the surrounding air, and [it comes about also] on account of certain others [of its] natural attributes that we will speak about later;[43]
but [responsible] for the lengths of times with regard to gestation
10   [is] the size of the [fetuses] that are being produced; for [it is] not
easy for big composites, whether consisting of animals or of any, so
to speak, of the other [beings], to attain their perfection in a short
time. Wherefore horses and the animals akin to these, even though
they live for a shorter time [than human beings], are pregnant for
a longer time; for the birth of the former [takes place] after a year,
while that of the latter [is] after ten months at most. And on account
15   of the same cause, the birth of elephants also [takes place] after a
long time; for their gestation [lasts] for two years on account of their
excess of size. Reasonably, for all [animals], the lengths of time of
their gestations and comings into being and their lives wish to be
measured in accord with nature by periods. I mean by "period" day
20   and night and month and year and the times that are measured by
these, and further, the periods of the moon. Periods of the moon
are full moon and its disappearance and the midpoints of the times
in between; for it is related to the sun in accord with these; for the
month is a period that belongs in common to both.[44] The moon is

---

43   This is a reference to the treatise *On Long-Livedness and Short-Livedness*, and
especially to its fifth chapter, which begins at 466a17. *On Long-Livedness and
Short-Livedness* is the next to last treatise among those now known as the *Parva
Naturalia*.

44   Since the moon gets its light from the sun, a lunar month (i.e., the time from
one full moon, or the moon's being in opposition to the sun, to another full
moon) depends on the motion of the sun as well as that of the moon itself.
And since the sun's yearly revolution is in the same direction as the moon's
monthly revolution, it follows that a lunar month is longer—by about two
days, as it happens—than the time it takes for the moon to orbit the earth.

a ruling beginning on account of its partnership with the sun and
its having a share in its light; for it comes into being like another,   25
lesser, sun; wherefore it contributes to all comings into being and
comings to perfection. For instances of heat and of cold, up to a
certain proportion, produce comings into being and, after these,
perishings. And the motions of these stars [i.e., of the sun and the   30
moon] fix the limit of these, of their beginning and of their end.
For just as we see both the sea and all the nature [that consists] of
what (pl.) is wet being still and changing in accord with the motion
and rest of the winds, and the air and the winds [doing so] in accord
with the period of the sun and the moon, so also [it is] necessary
for the things that spring forth out of these and that are in these to   35
follow; for [it is] in accord with reason that the periods of the less   778a
sovereign [beings] follow those of the more sovereign ones. For
there is a sort of lifetime, and a coming into being and a dying away,
even of the wind. And there might perhaps be some different ruling
beginnings of the revolution of these stars.[45] And so nature wishes
to number the comings into being and the endings [of the animals]   5
by the numbers of these, but it isn't exact on account of the indefi-
niteness of the material and on account of there coming into being
many ruling beginnings which, hindering the comings into being
and the perishings in accord with nature, are often responsible for
what (pl.) happens contrary to nature. And so it has been spoken   10
concerning the nourishing of the animals inside [the mother] and
concerning their coming into being outside, both separately concerning each [kind] and in common concerning them all.[46]

---

45 This is generally thought to be a reference to the Prime Mover and the other
Unmoved Movers that Aristotle argues for in the last book of the *Physics* and
in Book Lambda of the *Metaphysics*. Yet the tentativeness of his expression
here is worth remarking.

46 At the end of this sentence several of the more important manuscripts add
some words that Drossaart Lulofs, along with all the other modern editors I
am aware of, chooses not to include in his text. They would be translated
roughly as follows: "And [it has been spoken] concerning the differences by
which the parts of the animals differ, and this sort of thing happens especially
with regard to human beings." After these words, several of these manuscripts
also add another sentence fragment, which would be translated as "As many
parts, on the one hand, as all the animals have, both of the internal [ones] and
the external [ones]."

Book Five

Chapter 1

But the [incidental] attributes by which the parts of the animals differ must now be looked into. I mean such attributes of the parts as blueness and darkness of [the] eyes, and highness and lowness of [the] voice, and differences of [the body's] color[1] and of hairs or of feathers. Some of these sorts of [attributes] are in fact present in whole kinds, though in some [kinds they are present merely] as it happens; for example, this occurs especially in the case of human beings. Further, in connection with the changes belonging to the times of life, some [of these] are present equally in all the animals, whereas others are the opposite, like [the changes] with regard to voices and with regard to hair colors; for some [animals] do not turn noticeably gray toward old age, but the human being is affected in this respect more than the other animals. And some [of these attributes] follow immediately upon [the animals'] being born, whereas others become clear as they advance in age and grow old. Now concerning these and all such [attributes], one ought no longer to hold that there is the same manner of cause. For with regard to as many [things] as are neither products of nature in common nor peculiar to each kind, none of these either is or comes into being such [as it is] for the sake of something. For an eye is for the sake of something, but [it is] not blue for the sake of anything unless this attribute is peculiar to the kind. Nor in some [even of these] cases do they contribute to the defining char-

20

25

30

---

1   At 778a19, I follow Drossaart Lulofs, who deletes the words *ē sōmatos* after the
    word translated here as "color," even though they appear in all the manuscripts.
    The deletion has some support, however, from the Arabic translation, which
    gives no sign of them. If they were included, this whole phrase would be trans-
    lated: "and differences of color *either* of [the] body and of hairs *or* of feathers"
    (emphasis mine). But I can't see why Aristotle would present bodily color to-
    gether with hair color as an alternative category to the color of feathers.

35 acter (*logon*) of the [animal's] being, but on the assumption that they
778b come into being from necessity, their causes must be referred to the
material and the ruling beginning that initiated motion. For as was
said at the beginning in our first speeches,[2] [it is] not on account of
each [thing's] coming into being of a certain sort, [it is not] on account of this [that] it is of a certain sort, insofar as it concerns as
many [things] as are ordered and defined products of nature, but [it
is] rather on account of their being of these sorts [that] they come
5 into being of such sorts; for coming into being follows upon being
and is for the sake of being, but this doesn't [follow upon] coming
into being. But the ancients who spoke about nature supposed the
opposite; responsible for this [is] that they didn't see that the causes
are more [than one], but [saw] only the [cause consisting] of the
material and the [cause consisting] of motion, and these without dis-
10 tinguishing [them]; and they gave no consideration to the [cause
consisting] of the defining character and that [consisting] of the end.
So then each [thing] is for the sake of something, and it comes into
being, then, on account of this cause as well as on account of the remaining ones, insofar as it concerns as many [things] as are present
in the defining character of each thing, or are for the sake of something or [are that] for the sake of which. But for those [things] that
are not of these sorts, as many [of them] as have a coming into being,
15 when it comes to these, one ought to seek the cause in the motion
and the coming into being, on the assumption that they acquire their
differentiation in the process of formation itself. For [a given animal]
will have an eye for the sake of something (for it is presupposed that
it is an animal of such a sort [i.e., capable of seeing]), but [it will
have] an eye of such a sort from necessity, and not that sort of necessity, but another manner [of it, namely], that it is of a nature to
act and to be acted upon in this or that way. With these [things] hav-
20 ing been determined, let us speak about what (pl.) follows next. In
the first place, then, when the young of all [animals] are born, espe-
ially of those that bring them forth unperfected,[3] they are accus-

---

2 Cf. *On the Parts of Animals* 640a10ff. As is clear from the opening sentence,
On the Generation of Animals is a sequel to *On the Parts of Animals*, whether
it follows immediately or after a number of other treatises.

3 At 778b21, I accept Drossaart Lulofs' reading, *tōn atelē tiktontōn*, even though
it is supported only by the Arabic translation. The manuscripts read *tōn atelōn*,

tomed to sleeping, on account of their also passing the time sleeping inside the mother,[4] when they first acquire sense perception. But there is a perplexity with regard to their coming into being from the beginning, [namely,] whether awakeness or sleep belongs to animals earlier. For on account of their manifestly being awake more as they advance in age, [it is] reasonable that the opposite, [namely,] sleep, should belong to them at the beginning of their coming into being. Further, [this is reasonable] on account of the transition from not being to being coming about through what is in between; and sleep seems to be among such [things] in its nature, a boundary as it were between living and not living, and the one sleeping [seems] neither completely not to be nor [completely] to be. For living belongs most of all to being awake on account of sense perception. And if it is necessary that the animal have sense perception, and it is then first an animal when sense perception first comes into being, one ought to hold that its state at the beginning [is] not sleep but similar to sleep, like the [state] that the class of plants is also in. For in fact at this time animals live the life of a plant—but [it is] impossible for sleep to belong to plants; for [there is] no sleep from which there is no waking, and there is no waking from the condition of plants that is analogous to sleep—and so [it is] necessary for the [newly formed] animals to sleep for the greater [part of the] time on account of their growth and their weight being laid upon the upper regions (and we have said in other [speeches][5] that the cause of sleeping is of that sort); but nonetheless they manifestly wake up even in the womb (and this becomes clear in dissections and in the [case of the animals] that are brought forth as eggs); then they immediately fall asleep and drop off again. Wherefore also when they have come out [of the mother], they spend most of their time sleeping. Also, small children [at first] do not laugh when they are awake, but they both cry and laugh when they are sleeping. For sense per-

25

30

35

779a

5

10

---

which makes sense only if this phrase were to be translated as "especially of those [that are] unperfected [at birth]."

4 At 778b22, I read *mētri*, with all the manuscripts, instead of *mētrai*, with Drossaart Lulofs, whose reading is supported by the Arabic translation. According to Drossaart Lulofs' text, the translation of this phrase would be "inside the womb" instead of "inside the mother."

5 Cf. *On Sleep and Awakeness* 456b17–457a33; *On the Parts of Animals* 653a10–20.

ceptions[6] occur in animals even when they are asleep, not only what are called dreams, but even apart from the dream, as in those who

15　get up while they are asleep and do many [things] without dreaming. For there are some who get up while they are asleep and walk around, being able to see just like those who are awake. For they have sense perception of what (pl.) is happening without being awake, yet not like a dream. And it looks as if small children, being

20　as it were unskilled in being awake, on account of being accustomed [to it] have sense perception and live while they are sleeping. But as time goes on and their growth changes over to the lower [part of the body], they at that point wake up more and spend the greater [part of the] time in that way. But more than the other animals, they at first pass the time in sleep; for they are born the most unperfected among those that are [born] perfected, and with their growth most

25　of all in the upper part of the body. The eyes of all small children, immediately after having been born, are bluish, and later they change over to the nature that they are going to have; but in the case of the other animals this doesn't noticeably happen. Now respon-

30　sible for this [is] that the eyes of the others are more of one color, for example, cows are dark-eyed, the [eye] of all sheep is watery, of some [animals] the kind as a whole is gray-eyed or blue-eyed, and some are amber-eyed, as indeed [is] the greater number of goats[7] itself. But the eyes of human beings happen to be of many colors;

35　for [human beings] are blue-eyed and gray-eyed and some dark-
779b　eyed, and others amber-eyed. So that as for the others, just as [those of a single kind] don't differ from one another, so also they don't [differ] themselves from themselves [at different times], for it is not their nature to have more than one color. Of the other animals the horse, most of all, is many-colored; and indeed some of them come

5　to have eyes of different colors. Of the other animals, none is noticeably affected in this way, but some human beings come to have eyes of different colors. Now as for the other animals not changing noticeably [between] being young and [being] older, but this happening in the case of small children, one ought to suppose there to

---

6　For this extended use of the word translated as "sense perceptions" (*aisthēseis*), at least in relation to dreams, compare *On Dreams* 462a4.

7　At 779a33, the word translated as "amber-eyed" is *aigōpa*, whose literal meaning would be "goat-eyed."

be a sufficient cause, namely this, that this part is of a single color
in the ones but of many colors in the others. But responsible for the   10
[eyes] being bluish [in small children] and having no other color
[is] that the parts of the young are weaker, and being blue-eyed is a
sort of weakness. But one ought to have a grasp in general concerning the difference of eyes, on account of what cause some [animals]
are blue-eyed, some are gray-eyed, some are amber-eyed, and some
are dark-eyed. To take it to be the case that blue [eyes] are fiery, as   15
Empedocles asserts, and that dark ones have more water than fire,
and that on account of this the ones don't see keenly by day, the
blue ones, on account of a lack of water, whereas the others [don't]
at night on account of a lack of fire, is not said finely, if indeed one
must posit [the organ of] sight [to consist] not of fire but of water   20
in all [cases]. And further, though it is possible to give the cause of
the colors also in another manner, if indeed it is as it was said earlier
in the speeches concerning the senses and still earlier than these in
the [things] that have been determined concerning soul,[8] and [if we
were correct with regard to the fact] that this sense organ is [con-   25
stituted] of water, and [with regard to the] cause on account of
which [it is constituted] of water but not of air or of fire, one must
take this cause to be [responsible] for the things said. For some eyes
have more liquid, and some less, than [the amount that provides]
the suitable motion, while others have a suitable [amount]. And so
those eyes that have much liquid are dark on account of large
amounts [of it] not being easy to see through, while those that [have]   30
little [are] blue, as is manifest also in the case of the sea. For the
[part] of it that is easy to see through appears blue, that which is
less [easy to see through], watery, and that which is unfathomed on
account of depth, black and dark blue. And the eyes that are in between these then differ by the more and less. And one must suppose
the same cause to be [responsible] also for blue eyes not being keen-   35
sighted by day nor dark ones by night. For blue ones on account of   780a
their small amount of liquid are set in motion too much by light
and by what (pl.) is visible, [both] insofar as [the eye is] liquid and
insofar as [it is] transparent. But the motion of this part is seeing
insofar as [it is] transparent, not insofar as [it is] liquid. Dark eyes,

---

8   Cf. *On Sense Perception and Sense-Perceptibles* 438b5–20; *On Soul* 424b22–
    425a13.

5 on the other hand, on account of their great amount of liquid are
set in motion less. For nocturnal light [is] weak; at the same time,
liquid in general also becomes hard to set in motion at night. But
[the eye] ought neither to be unmoved nor [to be moved] more [in-
sofar as it is liquid] than insofar as it is transparent; for the stronger
10 motion drives out the weaker. Wherefore also when [people] change
from [looking at] strong colors, and when they go from the sun into
the dark, they don't see; for since the motion that is present in [the
eye] is strong, it hinders that from outside, and in general neither
strong nor weak sight is able to see bright [colors] on account of the
liquid being affected and being set in motion too much. Also, the
15 infirmities of each [sort of organ of] sight make [this] clear. For
cataract comes about more in the blue-eyed, and what are called
night-blindnesses in the dark-eyed. For the cataract is⁹ a sort of dryness of the eyes, wherefore also it occurs more in those who are getting old; for these parts too, like the rest of the body, become dry
20 toward old age; but night-blindness [is] an excess of liquid, wherefore it comes about more in those who are younger; for the brain of
these is more liquid. The [organ of] sight that is in between [having]
much and little liquid [is] best; for neither as being little [in liquid]
does it hinder the motion of the colors on account of being dis-
25 turbed nor on account of a large amount does it provide difficulty
of motion. [It is] not only the things that have been said [that are]
responsible for seeing dully or keenly, but also the nature of the skin
upon what is called the pupil; for it needs to be transparent, and [it
is] necessary that what is thin and white and smooth be of this sort,
30 thin so that the motion from outside may go straight, smooth so
that it not cast a shadow by becoming wrinkled (for [it is] also on
account of this [that] the old don't see keenly; for like the rest of
the skin, that of the eye becomes wrinkled as well as thicker as [peo-
ple] get old), and white on account of black not being transparent;
35 for black is this very [thing], that which is not seen through. Wherefore also lanterns are unable to illuminate if they are [made] out of

---

9   At 780a17–18, I follow Drossaart Lulofs, whose deletion of the word *mallon*,
     which appears in all the manuscripts, finds support in the Arabic translation.
     If *mallon* were read, the translation would be "is more a sort of dryness of the
     eyes" or "is rather a sort of dryness of the eyes" instead of "is a sort of dryness
     of the eyes."

skin of that sort. So then in old age and diseases [it is] on account
of these causes [that people] don't see keenly, whereas small children
appear blue-eyed at first on account of the small amount of liquid
[in their eyes]. [It is] mostly human beings and horses [that] come
to have eyes of different colors, on account of the same cause as that
on account of which only the human being turns gray and only the   5
horse among the other [animals] turns noticeably white-haired as
it gets old. For grayness is a sort of weakness of the liquid in the
brain and a lack of concoction, and [so is] blue-eyedness; for what
is excessively thin or excessively thick has the same effect, the one,
as a small amount of liquid, and the other, as a large amount. Thus
when nature is unable to make a match by either concocting or not   10
concocting the liquid in both eyes alike, but [concocts] this but not
that, then it results that [animals] come to have eyes of different
colors. As regards some animals being keen-sighted and others not,
there are two manners of cause. For keenness is spoken of in pretty
much two ways, and this is likewise the case with regard to hearing   15
and smelling. For to see keenly is spoken of in one sense as being
able to see what is far away, while in another [sense it is] to perceive
as distinctly as possible the differences among the [things] seen.
And these don't occur together in the same [beings]. For the same
person, using his hand to shade his eyes or looking through a tube,
will judge the differences of colors no more and no less [well], but   20
he will see further; those, at any rate, [who look up] out of pits and
wells sometimes even see stars. So that if one of the animals has a
large projection over its eye, but the liquid in the pupil [is] not pure
and not well suited to the motion from outside, and the skin on its   25
surface [is] not thin, this one will not be accurate with regard to the
differences of colors, but it will be able to see from far away, as if
from nearby,[10] more than those whose liquid and its covering are
pure, but who don't have a brow projecting over their eyes. For the
cause of seeing keenly in such a way as to perceive distinctly the dif-   30
ferences [of colors] is in the eye itself; for just as on a clean garment
even small stains become noticeable, so also in the [organ of] sight
that is pure even small motions [are] clear and produce sense per-

---

10 At 780b27, Drossaart Lulofs would delete the words *hōsper ei kai egguthen*,
    translated here as "as if from nearby," with support only from the Arabic
    translation, which gives no sign of them.

ception. But [it is] the position of the eyes [that is] responsible for
35   their seeing [things] far away and for the motion from distant visible
[objects] reaching [them]. For [animals] with protruding eyes don't
781a   see well from far away, while those that have eyes situated in a hollow
recess [are] capable of seeing [things] far away on account of the
motion not being dispersed into space but going straight. For it
doesn't make any difference whether one says that seeing, as some
assert, is by means of sight going forth [from the eyes] (for [on that
5   view] if there is nothing projecting over the eyes, [it is] a necessity
that [sight] be dispersed and that less [of it] fall upon the [things]
seen and that [things] far away be seen less well) or [whether one
says] that seeing is by means of the motion from the [things] seen;
for [there is] equally a necessity that sight see by means of motion.
And so [things] far away would be seen best if there were a continuous tube, as it were, from the [organ of] sight to the [thing] seen;
10   for the motion from the visible [objects] would not be dissipated;
but if not, [it is] a necessity for [things] far away to be seen more
accurately, the further [the "tube"] extends. And let there be these
causes for the difference in eyes.

Chapter 2

15   It is in the same manner also with regard to hearing and smelling;
for one [meaning] of hearing and smelling accurately is perceiving
as much as possible all the differences of the underlying [beings]
that are perceptible, and another is both hearing and smelling from
far away. And so [what is] responsible for judging the differences in
a fine way [is] the sense organ, as in the case of sight, if [the sense
20   organ] itself along with the membrane around it is pure. For the
channels of all the sense organs, as was said in the [speeches] concerning sense perception,[11] extend to the heart, and in those [ani-

---

11   Aristotle's *On Sense Perception and Sense-Perceptibles* does not say that the channels of all the sense organs extend to the heart, but only that the organs of taste
and touch are near the heart (439a1–2). But in *On Youth and Old Age and Life
and Death and Breathing*, a work that comes later in the series of treatises now
known as the *Parva Naturalia*, he does claim that while the organs of taste and
touch manifestly extend to the heart, it is necessary that those of the other
senses do so as well (469a5–14ff.). And since all the treatises in the *Parva Nat-*

mals] that don't have a heart, to what is analogous. And so the [chan-
nel of the organ] of hearing, since the sense organ is [constituted]
of air,[12] comes to an end where the innate breath causes pulsation in   25
some and in others breathing out and breathing in. Wherefore also
there comes to be understanding of the [things] said, so as to [be
able to] repeat what was heard; for as[13] [was] the motion [that] entered through the sense organ, of such a sort in return, as from
one and the same stamp, there comes to be the motion through the
voice so that one says what [one] has heard. And when yawning   30
and breathing out, [people] hear less than when breathing in, on
account of the ruling beginning of the sense organ of hearing being
[situated] upon the part that produces breathing[14] and its being
shaken and set in motion at the same time as the organ sets the
breath in motion; for the organ is set in motion while setting [the
breath] in motion. The same condition happens also in wet seasons   35
and climates,[15] and the ears seem to be filled with air (*pneumatos*)
on account of being close to the ruling beginning of the place of   781b
breathing (*pneumatikou*).[16] So then accuracy of judgment with regard to the differences both of sounds and of odors depends on the
sense organ and the membrane upon its surface being pure; for as
in the case of sight, so also with these sorts of [sense perceptions],

---

uralia are said explicitly to deal with themes connected with sense perception
(*On Sense Perception and Sense-Perceptibles* 436a5–b6), Aristotle's reference
here to his speeches "concerning sense perception" seems intelligible.

12 Cf. *On Soul* 425a3–5.

13 At 781a27, I read *hoia* with all the manuscripts (as I have learned from other
editions), instead of *hois*, which Drossaart Lulofs reads as a result of what
seems to be a simple mistake.

14 By "the part that produces breathing," Aristotle means the heart. Cf. *On Youth
and Old Age and Life and Death and Breathing* 479b17–19 and 480a16–24.

15 At 781a35, Drossaart Lulofs agrees with earlier modern editors who thought
that some words must be missing after the words translated here as "in wet
seasons and climates."

16 At 781b1, Drossaart Lulofs obelizes the words *tēi arkhēi tou pneumatikou topou*
and suggests instead *tēn arkhēn tōi pneumatikōi topōi*, a modern emendation
without any ancient authority. According to his suggestion, this entire phrase
would be translated as "on account of their ruling beginning being close to the
place of breathing," instead of "on account of being close to the ruling beginning of the place of breathing." If the manuscript reading is correct, the ears
referred to here would presumably include the passages that extend to the heart.

5   it results that all the motions are distinguishable.[17] And perceiving, but some [animals] not perceiving, from far away happens in a similar manner as with sight. For those [animals] that have as it were channels through the parts [concerned], [channels that project] far in front of the sense organs, these are able to perceive from far

10   away. Wherefore as many [animals] as have long nostrils, like the Laconian hounds, are keen-scented; for since the sense organ is inward [from these], the motions from far away are not scattered but go straight, as in the case of those who use their hands for shade in front of their eyes. Similarly also with all those whose ears [are] long and jut far out like a cornice, [ears] such as some of the quad-

15   rupeds have, and whose helical [channel] within [is] long; for these too, receiving the motion from far away, give it over to the sense organ. Now among the animals [the] human being has the least, so to speak, accuracy of the senses from far away, in relation to its size, but as for [accuracy] with regard to the differences, [it is] the

20   most keenly perceptive of all. Responsible [for this is] that its sense organ [is] pure and least earthy and body-like, and a human being has by nature the thinnest skin among the animals in relation to its size. Nature has also produced the [things] having to do with the seal in a reasonable way; for though it is a quadruped and brings forth [its young] alive, it doesn't have ears, but only chan-

25   nels. Responsible [for this is] that its manner of life is in water. For the part that is the ears is added to the [inner] channels with a view to preserving the motion of the air that is far away; thus it is not at all useful for it, but they would produce the opposite, admitting a large amount of water into themselves. And concerning sight and hearing and the sense of smell, it has been spoken.

---

17 Drossaart Lulofs encloses the entire passage from 781a20 ("For the channels of all the sense organs...") to 781b5 ("it results that all the motions are distinguishable") within double square brackets, indicating that in his view it is a later copyist's addition to the text, even if the material stems from Aristotle himself. I think, however, that it is a genuine part of the text, for besides its appearance in all the manuscripts, Aristotle will go on in the sequel to use the term *aisthētērion*, or sense organ (781b20, cf. b16), to refer, at least in part, to what this passage speaks of as "the ruling beginning of the sense organ of hearing" that is situated "upon the part that produces breathing" (781a31–32), a usage that would be nearly unintelligible if the passage were deleted.

Chapter 3

Hair is different in human beings both in relation to one another ac-   30
cording to their [different] ages and in relation to the classes of the
other animals, as many of them as have hair. And pretty much all
[animals] that produce animals inside themselves have it; for one
must take it that even the spiny ones among such [animals] have a
certain form of hair, for example, the [spines] of hedgehogs and if   35
there is any other such [animal] among those that produce animals   782a
[inside themselves]. There are differences of hair with respect to
stiffness and softness and with respect to length and shortness and
straightness and curliness and abundance and scantiness, and in addition to these also with respect to colors, with respect to whiteness
and blackness and the [colors] in between these. With regard to some   5
of these differences [animals] differ also according to their ages,
young or growing old, and this [is] especially noticeable in the case
of human beings. For they become hairier as they become older, and
some become bald on the front [parts] of the head. But they don't
become bald when they are boys, nor [do] women; but men [grow   10
bald] as they become advanced in age. And human beings become
gray on their heads as they grow old, but this doesn't come about
noticeably in any, so to speak, of the other animals, though [it does
so] in the horse most of all among the others. And human beings become bald on the front [parts] of the head, but they first become   15
gray on the temples. And no one becomes bald either on these or on
the back [parts] of the head. As for as many of the animals as don't
have hair, but what is analogous to it, for example, birds [have] feathers and the class of fish [has] scales, in these too some of these sorts
of attributes occur in accord with the same account. Now for the   20
sake of what [end] nature has made the class consisting of hair for
the animals has been said earlier in [our speeches about] the causes
concerning the parts of animals.[18] But when what [conditions] are
present, and on account of what necessities, each of these [varieties
of hair] results, it belongs to the present path of inquiry to make
clear. Now then the skin is chiefly responsible for [its] coarseness

---

18 Cf. *On the Parts of Animals* 658a18–19, where Aristotle says that hair is present, in those that have it, for the sake of protection.

25   and fineness; for in some [animals the skin] is thick and in others thin[19] and in some porous and in others compact. But further, the difference of the moisture that is in it is a contributory cause; for in some [animals] it is oily and in others watery. For in general the nature of the skin is assumed [to be] earthy; for since it is on the surface,

30   as the moisture evaporates it becomes solid and earthy. And hair and its analogue don't come into being from the flesh but from the skin, with its moisture evaporating and fuming up in them. Wherefore coarse [hair] comes into being from thick [skin] and fine [hair] from

35   thin skin. Now if the skin is more porous and thicker, [the hair] is coarse on account of the large amount of the earthy [substance] and

782b   on account of the large size of the pores; but if [it is] more compact, [the hair is] fine on account of the narrowness of the pores. And further, if the moisture is watery, since it dries up quickly, the hair doesn't get large in size, but if it is oily, the opposite [happens]; for

5   what is oily doesn't dry up easily. Wherefore in general those of the animals that have thicker skin have coarser hair, however not those [with the] most [thick skin] more [than the others], on account of the causes that have been stated, as for example the [less thick-skinned] class of pigs is related to that of cattle and to the elephant and to many of the other [animals]. Also on account of the same

10   cause, in human beings the hair on the head is coarsest; for this [is] the thickest [part] of the skin and [it is situated] over the most moisture, and further it is very porous. Responsible for the hair also being long or short[20] [is] that the evaporating moisture [is or] is not easily dried up. And of its not being easily dried up [there are] two causes,

15   how much [it is] and of what sort [it is]; for the moisture [is] not eas-

---

19 The Greek word translated in this sentence as "coarseness," in the case of hair, is the noun form of the word translated as "thick" in the case of skin, and the word translated as "fineness," in the case of hair, is the noun form of the word translated as "thin" in the case of skin. I have been translating these words with their usual translations, "thick" and "thin," but in the case of hair I prefer to translate them as "coarse[ness]" and "fine[ness]" because Aristotle is speaking of the circumference of each individual strand of hair, not the density of the strands of hair on the scalp.

20 At 782b12, Drossaart Lulofs would delete the words translated here as "or short," which are present in all the manuscripts. But the manuscript reading can be defended if the words "is or" are understood, and I have added them in brackets later in the sentence.

ily dried up if there is much [of it], and if [it is] oily. And on account
of this, the hair [that grows] from the head in human beings [is]
longest; for the brain, being wet and cold, provides a great abundance
of moisture. Straight hair and curly hair come about on account of
the rising fume in the hair. For if it is smoky, it is hot and dry and   20
makes the hair curly. For it is bent on account of being moved with
two motions; for what is earthy is moved downward and what is hot
upward. And since it is easily bent, it becomes twisted on account
of weakness; and this is curliness of hair. Well, it is possible to take
the cause in this way, but it is possible also that, on account of having   25
little moisture and much that is earthy, it is dried by the surrounding
[air] and shrivels up. For what is straight becomes bent if it suffers
evaporation, and hair contracts together as if being burnt over fire,
suggesting that its curliness is a shriveling up on account of a lack
of moisture through the agency of the heat of the surrounding [air].
A sign [of this is] that curly hair is stiffer than straight [hair]; for   30
what is dry [is] stiff. As many [animals] as have a lot of moisture [are]
straight-haired; for in this [sort of hair] the moisture advances by
flowing, not by dripping. And on account of this the Scythians on
the Black Sea and the Thracians have straight hair; for they themselves [are] moist and the surrounding air [is] moist; but the Ethiopi-   35
ans and those in hot [climates] have curly hair; for their brains and   783a
the air that surrounds [them are] dry. But some of the thick-skinned
[animals] have fine hair on account of the cause that was mentioned
earlier; for the finer the pores are, the finer [it is] necessary for the
hair to become. Wherefore the class of sheep has hair of that sort;   5
for wool is a multitude of hairs. But there are some among the animals that have hair which is soft, but less fine, as for example the
[class] of hares is in relation to that of sheep. For the hair of such
[animals] is on the surface of the skin. Wherefore it doesn't have
length, but it occurs [with] nearly the same [character] as scrapings   10
from linen [cloths]; for these too don't have any [great] length, but
are soft and don't accept weaving. The sheep in cold [climates] are
in the opposite condition to [that of] the human beings; for the
Scythians have soft hair, but the Sarmatian sheep have stiff hair. Responsible for this [is] the same [thing] as in the case of all wild [ani-   15
mals]. For the coldness solidifying [their hair] as a result of drying
[it] stiffens it; for as the heat is squeezed out, the moisture evaporates
along with it, and both the hair and the skin become earthy and stiff.

Responsible [for this] in the case of the wild [animals is] their living in the open, while for the others it is the region [where they live], which is of that [same] sort. A sign [of this is] also what happens in the case of sea urchins, which are used [as a remedy] for difficulty in urination. For these too, on account of being in sea [water] that is cold because of its depth (for they come to be [and are found] at sixty fathoms and even more), are themselves small, but their spines are big and stiff, big on account of the growth of their body having been diverted there (for on account of having little heat and not concocting their nutriment they have a lot of residue, and their spines and their hair and the [parts] of that sort come into being from residue), and stiff and petrified on account of the coldness and its congealing [effect]. And in the same manner also, in the case of the other [living beings, namely], the ones that spring forth [from the earth], those in north-facing [regions] are found to come into being stiffer and earthier and more stone-like than those [in regions] that face south, and those in windy [regions] than those in valleys; for they all are cooled more and their moisture evaporates off. Both the hot, then, and the cold cause stiffening; for the moisture ends up evaporating off through the agency of both, through that of the hot in itself and through that of the cold by concomitance (for it evaporates off together with the hot; for there is nothing wet without heat). But the cold not only stiffens but also condenses, whereas the hot makes [a substance] rarer. And also on account of the same cause, as [animals] become older the hair in those that have hair becomes stiffer, as do the feathers and scales in those that have feathers and scales. For their skins become harder[21] and thicker as they become older; for they become dry, and old age, in keeping with its name, is earthy[22] on account of the hot departing and along with it the wet. Human beings, most noticeably among the animals, go bald. But such a condition is something universal. For even among the plants some are evergreens while others shed their leaves, and among the birds, those

----

21 At 783b5, the Greek word that I translate here as "harder" with reference to skin is *sklērotera*, a form of the same word that I have been translating as "stiff" in the case of hair and spines and the like. Cf. n.19, above.

22 The Greek word for old age, *gēras*, is similar to *gē*, the word for earth, and *geēron*, the word translated as earthy, even though there is no etymological connection that I am aware of.

that hibernate shed their feathers. And baldness too is a condition of some such sort in human beings, in all those to whom it happens that they go bald; for there is a partial shedding of the leaves in all plants, and of the feathers and hair in those that have them, but when the condition comes about [with respect to these parts] all together, it acquires the names mentioned; for it is said that they go bald and that they shed their leaves.[23] Responsible for the condition [is] a lack of moisture that is hot, and [what is] especially of such a sort, among the moist [substances], is what is oily; wherefore also the oily ones among the plants [are] more evergreen. But with regard to these [beings], what is responsible must be stated in other [speeches]; for [there are] also other contributory causes of this condition in them. Now in plants the condition comes about in the winter (for this change has more influence than their age) and also for those of the animals that hibernate (for these too are less wet and hot in their nature than human beings); but human beings go through winter and summer in terms of their ages. Wherefore before [the age of] engaging in sex no one becomes bald; and then [it happens] more in those [inclined to] such [activity] in their nature. For by nature the brain is the coldest [part] of the body, and sexual activity cools; for it is an emission of pure natural heat. Reasonably, then, the brain feels[24] [the effect] first; for [things] that are weak and in a poor condition are [affected] by a small cause and added influence. So that if someone reckons that the brain itself has little heat, and further [that it is] necessary for the skin surrounding it to be of that sort [even] more and the nature of hair [even more] than this insofar as it is most distant, it would seem, reasonably, that going bald happens at about this age to those who are abundant in seed. And on account of the same cause human beings become bald only in the front [part] of the head, and only they among the animals [become bald], the front [part] because the brain is there, and only [they] among the animals

23 At 783b17, Drossaart Lulofs would add words meaning "and that they shed their feathers" after those translated here as "and that they shed their leaves." These words appear in an early printed edition of Aristotle, but they are present neither in the manuscripts nor in the oldest translations.

24 At 783b31, the word that I have translated as "feels" is *aisthanetai*, whose normal meaning is "to perceive [through the senses]" or "to have sense perception." Here, however, it is used in the sense of "to register" or "to show an effect." Cf. *On Dreams* 460a15 and n. 6, above.

because the human being has by far the largest and wettest brain.
5 And women do not go bald; for their nature is nearly the same as
that of small children; for both are unable to produce a spermatic
emission. And a eunuch doesn't become bald, on account of his
changing into the female [condition]. Also, as for the hair that orig-
inates later [in life], eunuchs either do not grow it or else they lose
it, if they happen to have it, except for pubic hair; for women too
10 don't have the former [sort of hair], but they do grow the [hair] that
covers the pubic region, and this deformation is a change from the
male to the female [condition]. As for the [animals] that hibernate
becoming hairy again and the [plants] that shed their leaves growing
leaves again, whereas [hair] doesn't grow again in those who are bald,
responsible [for this is] that for the former the seasons are turning
15 points of their body more [than in the case of human beings], so that
since these turn back, they turn back, both growing and losing, the
ones their feathers and their hair, and the plants their leaves. But for
human beings [it is] in terms of their age [that] there comes to be
winter and summer and spring and fall, so that since the ages don't
20 turn back, neither do the conditions turn back [that arise] on account
of these, despite the cause being similar. And the other conditions
of hair have pretty much been spoken about.

Chapter 4

Responsible for its colors in the other animals, and for their being
single-colored and variegated, [is] the nature of the skin; but in
25 human beings [this is responsible for] nothing [of the kind] except
for gray hair, not that on account of old age but that on account of
disease; for in the so-called white [pigment disease] the hair becomes
white, but if the hair is white <on account of old age>25 its whiteness
doesn't follow from the skin. Responsible [for these facts is] that it
grows out of the skin; therefore, out of skin that is diseased and white
30 the hair is diseased along with it, and disease of the hair is grayness.
But grayness of the hair on account of age comes about on account

---

25 At 784a27, Drossaart Lulofs encloses in angular brackets the words translated
here as "on account of old age," since they don't appear in the manuscripts,
but are supported only by the Arabic translation.

of weakness and a lack of heat. For all youthful prime tends toward
coldness as the body declines in old age; for old age is cold and dry.
And one must consider, with regard to the nutriment coming to each   35
part, that in each its own heat does the concocting, and when it is   784b
unable, [the part] perishes and there comes to be deformity or disease. (One must speak more precisely about this sort of cause in the
[speeches] concerning growth and nutriment.)[26] In all those human
beings, then, in whom the nature of their hair has little heat and the
moisture entering [it] is excessive, since its own heat is unable to con-   5
coct [it], it rots through the agency of the heat in the surrounding
[air]. And all rotting comes about through heat, but not that which
is innate, as has been said in other [speeches].[27] There is rotting of
water and of earth and of all such bodily [substances], wherefore also
of earthy vapor, for example, what is called mold; for mold is a rotting   10
of earthy vapor. So that also the nutriment in the hair, being of that
sort, rots if it is not concocted, and there comes to be what is called
grayness. And [it is] white because mold too, alone so to speak among
the [things] that are rotten, is white. Responsible for this [is] that it
contains much air; for all earthy vapor has the [same] power [as that]   15
of thick air. For mold is as it were the counterpart of hoarfrost, for
if the rising vapor freezes it becomes hoarfrost, whereas if it rots [it
becomes] mold. Wherefore also both are on the surface; for vapor is
on the surface. And so the poets in comedies use a good metaphor,
jokingly calling gray hair the mold and hoarfrost of old age. For the   20
one is the same in genus and the other in species, hoarfrost in genus
(for both are mist) and mold in species (for both are [forms of] rotting). A sign that [gray or white hair] is of such a sort: for gray hairs
have grown up in many [men] as a result of diseases, whereas later,
after they have become healthy, dark [hairs have grown up] instead   25
of these. Responsible [for this is] that in infirmity, just as the whole
body is in want of natural heat, so among its parts even those that
are very small share in this infirmity. And there comes to be much
residue in the bodies and in the parts; wherefore lack of concoction   30
in the flesh produces the gray hair. But after they have become
healthy and strong, they change back again and become young as it
were from [being] old. And it is right also to say that disease is an

---

26 No such treatise is known to exist.

27 Cf. *Meteorologica* 379a16–18.

acquired additional old age, and old age a natural disease; at all events some diseases produce the same [effects] as old age does. The tem-
35 ples go gray first. For the rear [parts] are empty of moisture on ac-
785a count of not containing [the] brain, whereas the bregma²⁸ contains much moisture, and that [of] which [there] is much does not rot eas-ily. But the hair on the temples has neither so little moisture as to be
5 concocted nor [so] much as not to rot; for its place, being in the mid-dle, is free from both conditions. And so what is responsible for the gray hair of human beings has been stated.

Chapter 5

But responsible for this change not noticeably coming about on ac-count of age in the other animals [is] the same [thing] as was stated in the case of baldness; for their brain is small and [so little] moist
10 that²⁹ its heat does not utterly lack the capacity for concoction. Of all the animals we know,³⁰ [this change] is most noticeable in horses because the bone they have around their brain is thinner in propor-tion to their size than in the others. Evidence [of this is] that a blow
15 to this place is fatal for them; wherefore also Homer composed [verses] as follows:

> where the first hairs grow on the skull of horses, and it is most fatal.³¹

---

28 Cf. 744a24 and Book Two, n. 53.

29 At 785a10, Drossaart Lulofs adds in angular brackets the word *hētton*, meaning "less," before the word translated here as "moist," though there is no ancient authority for this addition. According to his suggested text, the translation would read "less moist [than in man], so that" instead of "[so little] moist that."

30 At 785a11, I follow Drossaart Lulofs, who reads *tois d'hippois pantōn*, even though his reading is supported by no authority other than a printed text from the Renaissance. I have serious doubts as to whether this reading is correct, but since nothing important seems to be at stake, I am willing to accept it. Most of the manuscripts read *tois d'hippois autōn*, which is admittedly hard to construe, though it might possibly allow a translation of the beginning of this sentence as "Among them, of the animals we know."

31 Cf. *Iliad* 8.83–84.

Since, then, the moisture flows to these hairs easily on account of the thinness of the bone, and since their heat falls short on account of age, they turn gray. And red hair turns gray sooner than does dark; for redness is as it were an infirmity of hair, and all [things]   20
that are weak grow old more quickly. Now it is said that cranes become darker as they grow old. Responsible for their being affected [in this way] would be that the nature of their feathers is by nature rather white,[32] and that as they grow old the moisture in their feathers is too much to be easily rotted.[33] That gray hair comes to be by   25
a sort of rotting, and that it is not, as some suppose, a drying up, a sign of what was said before[34] [is] that hair covered by caps or veils turns gray more quickly (for winds hinder rotting, and the covering produces a lack of wind) and that an ointment of water and oil   30
mixed together helps [against it]. For the water cools [the hair], while the oil mixed in prevents [it] from drying quickly; for water [is] easily dried. And that it is not a drying up, and that hair is not whitened in the way that grass [is], by drying up, a sign [is] that some hairs grow gray straightaway; but nothing [that is] dry grows.   35
But many hairs become white also at the tip; for in the extremities and thinnest [parts] there comes into being the least [amount of] heat. In the other animals, as many as in which white hair comes   785b
into being, this happens to come about by nature, but not because of a [diseased] affection. Responsible for the colors in the others is their skin. For the skin of those [whose hair is] white is manifestly white, of those [whose hair is dark], dark, and of those [whose hair is] variegated and comes to be from a mixture, white in one part   5
and dark in another. But in the case of human beings the skin [is] not at all responsible; for even those who are white have extremely dark hair. Responsible [for this is] that the human being has the

---

32 At 785a23, I read *leukoteran*, with all the ancient authorities, instead of *leptoteran*, a modern emendation accepted by Drossaart Lulofs. If one were to accept that reading, the translation would read "rather thin," instead of "rather white."

33 At 785a25, I read *eusēpton*, with Drossaart Lulofs, even though it has no ancient authority except for apparent support from the Arabic translation. The reading of all the manuscripts is *eusēptoteron*, which would lead to the translation "rather easily rotted," or "more easily rotted," instead of "easily rotted."

34 At 785a27, Drossaart Lulofs would delete the words translated here as "of what has been said before," with apparent support from the Arabic translation, even though they appear in all the manuscripts.

thinnest skin of all in proportion to its size, wherefore it doesn't
10   have the strength for [producing] a change of the hair, but the skin
itself changes in color on account of its weakness and becomes
darker as a result of sunny days and winds; and the hair doesn't
change at all along with it. But in the other [animals] the skin, on
account of its thickness, has the power of a [surrounding] region;[35]
wherefore their hair changes in accord with their skin but their
15   skins don't [change] at all in accord with the winds and the sun.

Chapter 6

Some of the animals are single-colored (I mean by "single-colored"
those whose kind as a whole has one color, for example, all lions are
tawny, and likewise this is [found] also in birds and in fish and the
20   other animals), some are many-colored but whole-colored (I mean
those whose body as a whole has the same color, for example, an ox
is white as a whole or black as a whole), and some are variegated,
and this in two ways, some in their kind, as leopard and peacock
and some of the fish, for example, the so-called *thrattai*, while in
the case of others the kind [is] not variegated in its entirety, but var-
25   iegated ones do come into being, for example, oxen and goats, and
among birds, pigeons, for example; and the same [thing] happens
to other kinds of birds. Those that are whole-colored change much
more than the single-colored, both into one another's simple color,
for example black from white [parents] and white from black ones,
30   and into [colors that are] mixed from both, on account of its being
in their nature not to have one color for the whole kind; for the kind
is easily moved in both directions so that they both change into one
another and are also variegated more. Those that are single-colored
[are] the opposite; for they don't change except on account of something happening [to them], and this [occurs] rarely; [but it does
35   occur,] for a white partridge has in fact been seen, and a crow and a
sparrow and a bear. These things happen when they have been dis-
786a   torted in the [process of] coming into being. For what is small is
easily corrupted and easily moved, and what is coming into being
is of that sort; for the ruling beginning in [beings] that are coming

---

35 Cf. 767a28–32.

into being is in [something] small. Those [that] change most of all [are those] that, while being single-colored³⁶ by nature, are many-colored when taken as the kind, on account of the waters [they drink]; for hot [waters] make their hair white, whereas cold [waters make it] black, just as also in the case of plants. Responsible [for this is] that hot [waters] contain more air than water, and the air shining through produces whiteness, as also in foam. And so, just as skins that are white on account of something happening to them [differ] from those [that are white] on account of nature, also in the case of hair the whiteness of the hair on account of disease and age differs from that on account of nature, because of the responsible [agent] being different. For natural heat makes the latter white, whereas external heat [makes] the former [white]. And the vaporous air enclosed within provides the whiteness in all [cases]. Wherefore also, as many [animals] as are not single-colored are all whiter in the [parts] under the belly. And indeed all white [animals] so to speak are hotter and have better-tasting flesh on account of the same cause; for concoction makes them sweet, and heat [produces] concoction. And the same cause [is] also [responsible] for those that, [being] single-colored, are [either] black or white; for heat and coldness is responsible for the nature of skin and of hair; for each of the parts has its own heat. Further, the tongues differ, [those] of the simple[-colored] and variegated [animals] and [those] of the [animals] that are simple[-colored] but differing [in color], for example, white and black. Responsible [for this is] what was said earlier, that the skins of those that are variegated are variegated and [the skins] of those that are white-haired and black-haired are, the former, white, and the latter, black. And one must take the tongue to be, as it were, one part among the external [ones], regardless of the fact that it is covered inside the mouth, but [to be] like a hand or a foot; so that since the skin of those that are variegated is not single-colored, this [is] responsible also for the skin on the tongue [not being

36 At 786a3, I read *monokhroa*, with the great preponderance of the manuscripts, rather than *holokhroa*, which Drossaart Lulofs reads even though it appears only in a single, rather late manuscript. According to Drossaart Lulofs' text, the translation would read "whole-colored" instead of my "single-colored." Drossaart Lulofs' reading is admittedly more consistent with Aristotle's use of the terms at the beginning of this chapter, but Aristotle may be deliberately modifying that earlier usage, as he seems certainly to be doing at 786a14, below.

30   so]. Some among the birds and some among the wild quadrupeds change their colors in accord with the seasons. Responsible [for this is] that, as human beings change in accord with their age, this happens to those in accord with the seasons; for this difference is greater [in their case] than the change in accord with age. And also, the more

35   omnivorous [animals] are more variegated, to speak for the most

786b   part, reasonably [so], for example, bees are more single-colored than hornets and wasps; for if nutriments are responsible for the change, varied nutriments reasonably make the motions and the residues of the nutriment more manifold, out of which hair and feathers and

5   skins come into being. And concerning colors and hair let it have been determined in this manner.

Chapter 7

With regard to the voice, that some of the animals are low-voiced, others high-voiced, and others well-pitched and suitably related

10   to both extremes, and further [that] some are big-voiced and others small-voiced and [that] they differ from one another in smoothness and roughness and flexibility and inflexibility [of the voice], one must examine on account of what causes each of these [attributes] belongs [to it]. Now then with regard to highness and lowness one must suppose that there is the same cause as in the case of the change which [animals] undergo [in first] being young and [then]

15   older. For all the other [animals] call with a higher [voice] when they are younger, but among cattle the calves [call with] a lower [voice]. The same [thing] happens also in the case of males and females. For in the other kinds the female calls with a higher [voice] than the male (this is most noticeable in the case of human beings;

20   for nature has given this capacity most of all to these on account of their using speech, alone among the animals, and the voice being the material of speech), but in the case of cattle [it is] the opposite; for the females call with a lower [voice] than the bulls. Now that for the sake of which animals have a voice, and what is voice and sound in general, some [things] have been said in the

25   [speeches] concerning sense perception[37] and others in those con-

---

37 *On Sense Perception and Sense-Perceptibles* 446b29–447a1.

cerning soul.[38] But since low consists in the motion being slow and high in its [being] fast, whether that which sets in motion or that which is moved [is] responsible for [its being moved] slowly or quickly involves a certain perplexity. For some assert that what is much is moved slowly and what is little quickly, and that this is [the]   30
cause of some [animals] being low-voiced and others [being] high-voiced, speaking finely up to a certain point, but as a whole not finely. For in terms of the genus it seems as if it is spoken correctly that lowness consists in a certain magnitude of what is moved. [but it is not spoken correctly.] For if this [is one's definition], [it is] not easy to [explain how animals] call with a [voice that is] both small and low, nor similarly with one [that is] big[39] and high. Also, low-ness of voice is thought to be characteristic of a nobler nature, and   35
in songs what is low [is] better than what (pl.) is high-pitched; for   787a
what is better consists in preeminence, and lowness is a sort of pre-eminence. But since the low and the high in voice [is] different from bigness of voice and smallness of voice (for there are high-voiced, big-voiced [animals] and likewise small-voiced, low-voiced   5
ones) and similarly for the pitch intermediate between these, by what else would someone give a definition concerning these (I mean bigness of voice and smallness of voice) than by a big or small amount of what is moved? If then the high and low will be in accord with the definition stated [by these people], it will turn out that the same [animals] are [always] low-voiced and big-voiced, and high-   10
voiced and small-voiced. But this [is] false. Responsible [for this per-plexity is] that the [terms] big and small and much and little are in some cases spoken unqualifiedly but in others relatively to one an-other. So that [for animals to be] big-voiced consists in that which is moved being much, and [for them to be] small-voiced in [its being] little, unqualifiedly, but [for them to be] low-voiced and high-voiced [consists] in [that which sets in motion and that which is moved] having this difference in relation to one another. For if that   15
which is moved surpasses the strength of that which sets in motion,

---

38 *On Soul*, Book Two, Chapter 8, and especially 420b5–421a6.

39 At 786b34, I follow Drossaart Lulofs in reading *mega*, meaning "big," even though almost all the manuscript authority is in favor of the reading *baru*, meaning "low." Drossaart Lulofs' text is supported however by the Arabic translation, and perhaps by an erased and overwritten word in manuscript Z.

[it is] a necessity that what is (locally) moved be in (local) motion slowly, but if it is surpassed, quickly. And that which is strong, on account of its strength sometimes setting much in motion, produces a motion that is slow, but sometimes, on account of over-

20   powering [it], [produces one that is] fast. And according to the same account, some of weak ones among the movers, setting in motion more than their power [can easily move], produce a motion that is slow, but others of them, on account of weakness setting little in motion, [produce one that is] fast. So then these [are] the causes of the contrarieties: of neither all the young being high-

25   voiced nor low-voiced, nor [all] those that are older, nor [all] the males and females, and in addition to these [things], of the sick, as well as those in a good condition of the body, calling with a high [voice], and further also, of [men] as they become old becoming more high-voiced, even though their age is opposite to that of the young. So then most [animals] when they are younger and [most] females, on account of [their] lack of power setting little air in mo-

30   tion, are high-voiced; for the small amount [of air] is in fast (local) motion, and in the voice, what is fast [is] high. But calves and fe-male cattle, though the part by which they produce motion is lack-ing in strength, in the former on account of age and in the latter on account of the nature of femaleness, set a large amount in mo-

787b   tion and thus call in a low [voice]; for what is in (local) motion slowly [is] low, and the large amount of air is in (local) motion slowly. These set much in motion, but the others little, because in these the cavity through which the breath is first moved (locally) has a large extent and they are compelled to set much air in motion,

5   whereas in the others it is easily dispensed. But as their age ad-vances, this part, the one that sets in motion, becomes stronger in each, so that they change into the opposite, and the high-voiced become lower-voiced, themselves in relation to themselves, and the low-voiced higher-voiced; wherefore bulls [are] higher-voiced

10   than calves and female cattle. Now then in all [animals] strength is in the sinews, wherefore also those in their prime are stronger; for the young are more unarticulated and lacking in sinews. Further, in the young tension is not yet established, whereas in those that have grown old it is already relaxed; wherefore both are weak and

15   without power for [producing] motion. Most sinewy [are] bulls, and [thus so is] their heart; wherefore this part in them, by which they

210

set the breath in motion, is taut, like a stretched sinewy cord. And
it is made clear that the heart of cattle is of such a sort in its nature
by [the fact that] a bone comes into being in some of them; for
bones seek [to have] the nature of sinew. All [animals] when cas-   20
trated change to the female [condition], and on account of their
sinewy strength being relaxed in its ruling beginning, they produce
a voice like that of females. The relaxation that comes about is
nearly the same as if someone had stretched a cord and made it
taut by hanging some weight [from it], as the [women] who weave
at looms do; for these stretch the warp by attaching what are called   25
loom-stones. For in this way also the nature of the testicles is attached to the spermatic channels, while these [come] from the
blood vessel whose ruling beginning is from the heart, near the
very [part] that sets the voice in motion. Wherefore also when the
spermatic channels change at about the age when [the young] are   30
now able to secrete seed, this part changes along with them. And
when this part changes, the voice also changes, more so in males,
but the same [thing] occurs also in the case of females, though less
evidently, and there comes to be what some call "bleating" [like a   788a
goat], when the voice is uneven. But afterwards, [the voice] settles
into the lowness or highness of the age that follows. But when the
testicles are removed, the stretching of the channels is relaxed, just
as when the weight is removed from the cord and the warp. And   5
when this is relaxed, the ruling beginning that sets the voice in motion is also loosened in the same ratio. So then on account of this
cause those [animals] that are castrated change to the female [con-
dition], in voice and in the rest of their shape, on account of its following [from the castration] that the ruling beginning from which
tension is present in the body is relaxed, but not as some take it,   10
that the testicles themselves are a knot of many ruling beginnings;
rather, small changes become causes of great [things], not through
themselves, but when it follows that a ruling beginning changes
along with them. For ruling beginnings, though they are small in
size, are great in power; for this is [what it is for something] to be
a ruling beginning, for it to be a cause of many [things], but noth-   15
ing higher [to be a cause] of it. The heat and the coldness of the
region [where they live] also contribute to some among the animals
being constituted by nature in such a way as to be low-voiced and
others high-voiced. For hot breath on account of its thickness pro-

duces lowness of voice, while cold [breath] on account of its thin-
20   ness [produces] the opposite. This [is] clear also in the case of musical pipes; for those who have a hotter breath and blow in such a
manner as those who groan, play with a lower [pitch]. Responsible
for roughness of voice and for the voice being smooth and for all
that sort of unevenness [is] that the part and the organ through
25   which the voice is carried is either rough or smooth or in general
even or uneven ([this is] clear whenever some wetness is present
around the windpipe or there comes to be roughness as a result of
something happening to it; for then the voice too becomes uneven).
[Responsible] for flexibility[40] [is] if the organ is soft or hard; for
30   what is soft is able to be regulated and to become of every sort, but
what is hard is not able. And the soft [organ] is able to call with
both a small and a big [voice], wherefore also a high and low one;
for it easily regulates the amount of breath, since it also easily becomes big and small itself. But hardness can't regulate. So then
788b   concerning voice, with regard to as many [things] as have not been
determined earlier in the [speeches] concerning sense perception
and in those concerning soul,[41] let this much (pl.) have been stated.

Chapter 8

Concerning teeth, it has been said earlier[42] that animals don't have
them for the sake of one [thing] and that they don't all have them
5   for the sake of the same [thing], but some on account of nourishment
and others also for the sake of fighting and with a view to speech
[embodied] in the voice; but as for why the front [teeth] come into
being earlier and the molars later, and the latter don't fall out but
the former fall out and grow back again, one must hold that the cause
belongs to the genus of the speeches concerning coming into being.
10   Now then Democritus too has spoken about them, but he has not
spoken finely; for without having investigated with respect to all

---

40 At 788a28, Drossaart Lulofs would add the words *kai tēs akampsias*, meaning
    "and inflexibility" after the word translated as "flexibility," even though there
    is no ancient authority for this addition.

41 In other words, in *On Sense Perception and Sense-Perceptibles* and in *On Soul.*

42 Cf. *On the Parts of Animals* 655b8–11; 661a34–b27.

[cases] he states the cause universally. For he asserts that they fall out on account of their coming into being in animals before their time; for when [the animals] are in their prime so to speak, [it is then, he asserts, that] they grow according to nature, and he gives suckling as the cause for their coming into being before their time. And yet the pig also suckles, but it doesn't shed its teeth; and further, all the saw-toothed [animals] suckle, but some of them, for example, lions, don't shed [any] except for the canines. Thus he erred in this by speaking universally without having investigated what happens in all [cases]. But one must do this; for [it is] a necessity that the one who speaks universally say something concerning all [cases]. Now since we assume [with regard to] nature, making our assumption on the basis of what (pl.) we see, that it neither omits nor produces in vain any of the [things] possible with regard to each [being], and [since it is] a necessity for those [beings] that are going to take nutriment, after the enjoyment of milk [has ceased],[43] to have organs for working on their nutriment,[44] if, then, [the coming into being of teeth according to nature] happened as that one says, at around puberty, nature would be omitting something among the [things] possible for it to produce, and the work of nature would come about contrary to nature. For what is by violence is contrary to nature, and he asserts that the coming into being of the [first] teeth happens by violence. So then from these and other such [things], [it is] manifest that this [is] not true. These [teeth, i.e., the front ones,] come into being earlier than the flat ones, first because the work of these [is] earlier (for dividing is earlier than grinding, and those are for grinding and these for dividing), and next because what is smaller naturally comes into being more quickly than what is larger, even if they start out together. And these are smaller in size than the molars also because of[45] the jawbone being flat there but narrow near the [open-

15

20

25

30

789a

---

43 At 788b23–24, I read *meta tēn tou galaktos apolausin* with the manuscripts, rather than *meta tēn apogalaktisin* with Drossaart Lulofs, who has support only from William of Moerbeke's Latin translation. According to Drossaart Lulofs' text, the translation would read "after weaning," instead of "after the enjoyment of milk [has ceased]."

44 An explicit conclusion from the initial premises in this sentence is lacking also in the Greek.

45 At 789a1, I read *kai tōi to ostoun*, which appears (though without the iota in the word *toi*) in one relatively late manuscript. Drossaart Lulofs would further

ing of the] mouth. Accordingly, [it is] necessary that more nutriment
flow from what is larger, and less from what is narrower.[46] Suckling
5   itself contributes nothing, but the heat of the milk makes the teeth
sprout more quickly. A sign [of this is] that even among those that
suckle [considered by] themselves, those of their young that enjoy
hotter milk put forth teeth more quickly; for heat promotes growth.
[The front teeth] fall out after having come into being, on the one
hand for the sake of what is better, because what is sharp is quickly
10   blunted; thus others have to take their place with a view to their work.
But there is no [such thing as] bluntness of the flat [teeth], but they
are only made smooth from being worn down by time. But on the
other hand, [the front teeth] fall out from necessity, because the roots
of those [others] are in the flat [part of the] jaw and in a strong bone,
whereas [those] of the front [teeth] are in a thin [part of the jaw],
wherefore they are weak and easily moved. But they grow back again
15   because their shedding comes about in the bone while it is still growing and while there is still time for teeth to come into being. A sign
of this [is] that also the flat ones continue to grow for a long time;
for the last ones spring up at around twenty years [of age], and in
some [cases] in fact the last ones have come into being when they
were already quite old, on account of there being much nutriment
20   in the wide part of the bone. But the front [part], on account of its
789b   thinness, quickly reaches the end [of its growth], and there doesn't
come to be residue in it, but its nutriment is expended on its own
growth. Democritus, however, having neglected to speak of the for
the sake of which, ascribes to necessity all the [things] that nature
employs, which though they are of that sort, are nonetheless for the
5   sake of something and for the sake of what is better with regard to
each [being]. So that nothing prevents [teeth] from coming into
being and falling out in that way, yet not on account of these [things],

---

delete the word *kai* from this reading, even though the deletion has no ancient
authority. His reading would lead to the translation "because of " instead of
my "also because of." The preponderant manuscript reading is *kai to ostoun*,
which I cannot construe and which I wouldn't be able to translate.

46   At 789a3–4, I accept Drossaart Lulofs' reading of *ek de tou stenōterou elattō*,
even though its only ancient authority is the Arabic translation. The manuscripts read *ek de tou elattonos stenōteran*, which would require the unintelligible translation "and narrower [nutriment] from what is less," instead of "and
less from what is narrower."

but on account of the end; these [things are] responsible as setting in motion and as instruments and as material, since indeed [it is] plausible that most [things] are produced by means of breath as an instrument; for just as some of the [instruments] in the case of the   10 arts have many uses, like the hammer and the anvil in the smith's art, so also [does] breath in the [beings] constituted by nature. But to say that the responsible [factors are] from necessity looks the same as if someone were to suppose that the water has been drawn off from those with edema only on account of the knife, but not on account of being healthy, for the sake of which the knife made the cut. So   15 then concerning teeth, why some fall out and come into being again, whereas others don't, and in general on account of what cause they come into being, has been stated. And [our account] has been stated also concerning the other attributes having to do with the parts, as many [attributes] as happen to come into being not for the sake of something, but from necessity and on account of the cause that sets   20 in motion.